AN ASTRONAUT'S

GUIDE TO LIFE

ON EARTH

*

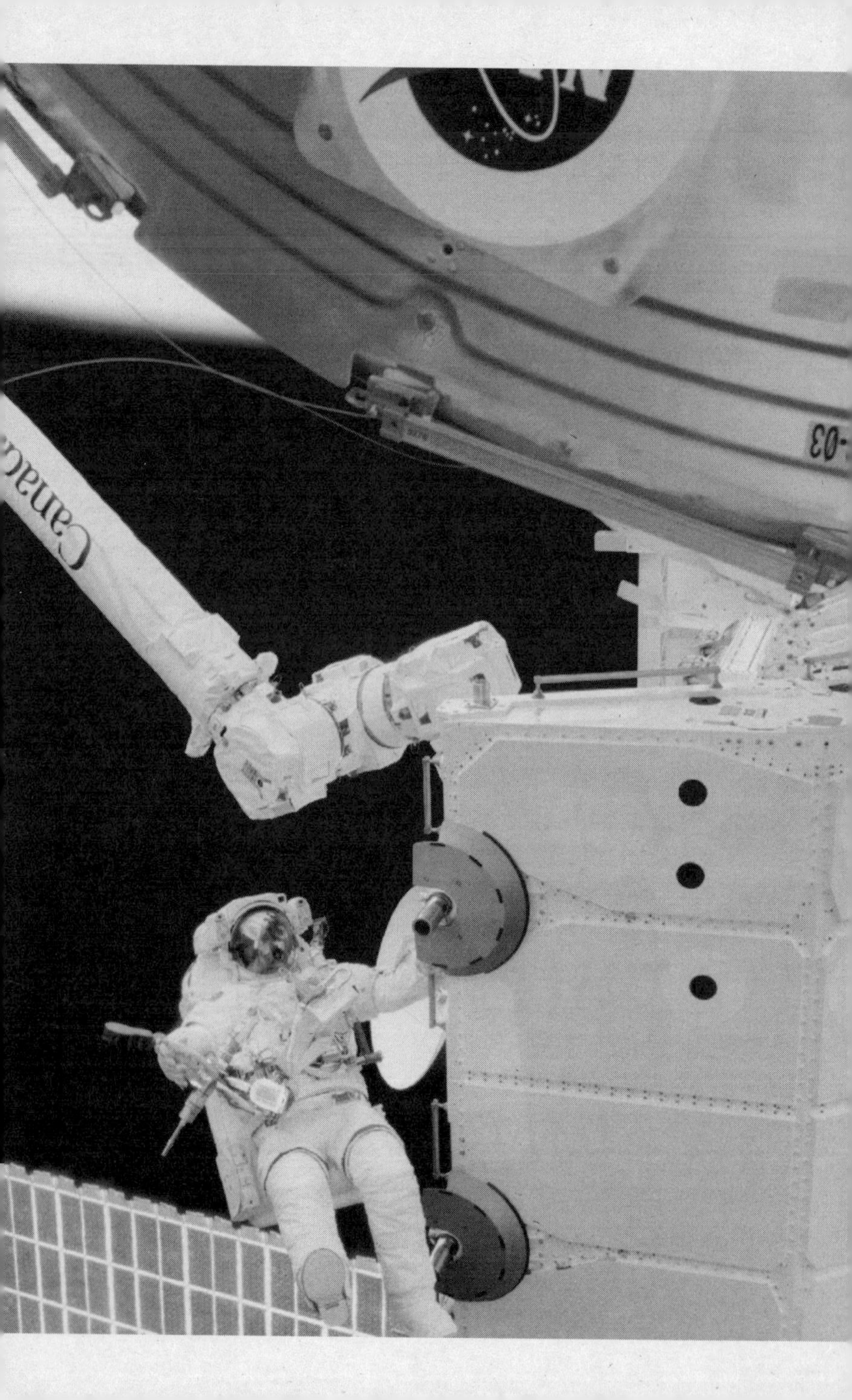

AN ASTRONAUT'S GUIDE TO LIFE ON EARTH

CHRIS HADFIELD

PAN BOOKS

First published 2013 by Macmillan

This paperback edition published 2015 by Pan Books
an imprint of Pan Macmillan
The Smithson, 6 Briset Street, London EC1M 5NR
Associated companies throughout the world
www.panmacmillan.com

First published simultaneously 2013 in Canada by Random House of Canada Ltd
and in the US by Little, Brown and Company

ISBN 978-1-4472-5994-7

Interior image credits: p. ii, Chris Hadfield on mission STS-100 spacewalk, credit: NASA;
pp. 20–21, cyclone off the African coast, credit: NASA / Chris Hadfield; pp. 136–37, moonrise,
credit: NASA / Chris Hadfield; pp. 240–41, Soyuz landing, credit: NASA / Carla Cioffi.

19

A CIP catalogue record for this book is available from the British Library.

Printed and bound by CPI Group (UK) Ltd, Croydon, CR0 4YY

To Helene, with love.
Your confidence, impetus and endless help
made these dreams come true.

CONTENTS

INTRODUCTION

MISSION IMPOSSIBLE

THE WINDOWS OF A SPACESHIP casually frame miracles. Every 92 minutes, another sunrise: a layer cake that starts with orange, then a thick wedge of blue, then the richest, darkest icing decorated with stars. The secret patterns of our planet are revealed: mountains bump up rudely from orderly plains, forests are green gashes edged with snow, rivers glint in the sunlight, twisting and turning like silvery worms. Continents splay themselves out whole, surrounded by islands sprinkled across the sea like delicate shards of shattered eggshells.

Floating in the airlock before my first spacewalk, I knew I was on the verge of even rarer beauty. To drift outside, fully immersed in the spectacle of the universe while holding onto a spaceship orbiting Earth at 17,500 miles per hour—it was a moment I'd been dreaming of and working toward most of my life. But poised on the edge of the sublime, I faced a somewhat ridiculous dilemma: How best to get out there? The hatch was small and circular, but with all my tools strapped to my chest and a huge pack of oxygen tanks and electronics strapped onto my back, I was square. Square astronaut, round hole.

The cinematic moment I'd envisioned when I first became an astronaut, the one where the soundtrack swelled while I elegantly pushed off into the jet-black ink of infinite space, would not be happening. Instead, I'd have to wiggle out awkwardly and patiently, focused less on the magical than the mundane: trying to avoid snagging my spacesuit or getting snarled in my tether and presenting myself to the universe trussed up like a roped calf.

Gingerly, I pushed myself out headfirst to see the world in a way only a few dozen humans have, wearing a sturdy jetpack with its own thrusting system and joystick so that if all else failed, I could fire my thrusters, powered by a pressurized tank of nitrogen, and steer back to safety. A pinnacle of experience, an unexpected path.

Square astronaut, round hole. It's the story of my life, really: trying to figure out how to get where I want to go when just getting out the door seems impossible. On paper, my career trajectory looks preordained: engineer, fighter pilot, test pilot, astronaut. Typical path for someone in this line of work, straight as a ruler. But that's not how it really was. There were hairpin curves and dead ends all the way along. I wasn't destined to be an astronaut. I had to turn myself into one.

* * *

I started when I was 9 years old and my family was spending the summer at our cottage on Stag Island in Ontario. My dad, an airline pilot, was mostly away, flying, but my mom was there, reading in the cool shade of a tall oak whenever she wasn't chasing after the five of us. My older brother, Dave, and I were in constant motion, water-skiing in the mornings, dodging chores and sneaking off to canoe and swim in the afternoons. We didn't have a television set but our neighbors did, and very late on the evening of July 20, 1969, we traipsed across the clearing between our cottages and jammed ourselves into their living room along with just about everybody else on the island. Dave and I perched on the back of a sofa and craned our necks to see the screen. Slowly, methodically, a man descended the leg of a spaceship and carefully stepped onto the surface of the Moon. The image was grainy, but I knew exactly what we were seeing: the impossible, made possible. The room erupted in amazement. The adults shook hands, the kids yelped and whooped. Somehow, we felt as if we were up there with Neil Armstrong, changing the world.

Later, walking back to our cottage, I looked up at the Moon. It was no longer a distant, unknowable orb but a place where people walked, talked, worked and even slept. At that moment, I knew what I wanted to do with my life. I was going to follow in the footsteps so boldly imprinted just moments before. Roaring around in a rocket, exploring space, pushing the boundaries of knowledge and human capability—I knew, with absolute clarity, that I wanted to be an astronaut.

I also knew, as did every kid in Canada, that it was impossible. Astronauts were American. NASA only accepted applications from U.S. citizens, and Canada didn't even have a space agency. But . . . just the day before, it had been impossible to walk on the Moon. Neil Armstrong hadn't let that stop him. Maybe someday it would be possible for me to go too, and if that day ever came, I wanted to be ready.

I was old enough to understand that getting ready wasn't simply a matter of playing "space mission" with my brothers in our bunk beds, underneath a big *National Geographic* poster of the Moon. But there was no program I could enroll in, no manual I could read, no one even to ask. There was only one option, I decided. I had to imagine what an astronaut might do if he were 9 years old, then do the exact same thing. I could get started immediately. Would an astronaut eat his vegetables or have potato chips instead? Sleep in late or get up early to read a book?

I didn't announce to my parents or my brothers and sisters that I wanted to be an astronaut. That would've elicited approximately the same reaction as announcing that I wanted to be a movie star. But from that night forward, my dream provided direction to my life. I recognized even as a 9-year-old that I had a lot of choices and my decisions mattered. What I did each day would determine the kind of person I'd become.

I'd always enjoyed school, but when fall came, I threw myself into it with a new sense of purpose. I was in an enrichment program that year and the next, where we were taught to think more critically and analytically, to question rather than simply try to get the right answers. We memorized Robert Service poems, rattled off the French alphabet as quickly as we could, solved mind-bending puzzles, mock-played the stock market (I bought shares in a seed company on a hunch—not a profitable one, it turned out). Really, we learned how to learn.

It's not difficult to make yourself work hard when you want something the way I wanted to be an astronaut, but it sure helps to grow up on a corn farm. When I was 7 years old we'd moved from Sarnia to Milton, not all that far from the Toronto airport my dad flew in and out of, and my parents bought a farm. Both of them had grown up on farms and viewed the downtime in a pilot's schedule as a wonderful opportunity to work themselves to the bone while carrying on the family tradition. Between working the land and looking after five kids, they were far too busy to hover over any of us. They simply expected that if we really wanted something, we'd push ourselves accordingly—after we'd finished our chores.

That we were responsible for the consequences of our own actions was just a given. One day in my early teens, I drove up a hedgerow with our tractor a little too confidently—showing off to myself, basically. Just when I got to feeling I was about the best tractor driver around, I hooked the drawbar behind the tractor on a fence post, breaking the bar. I was furious at myself and embarrassed, but my father wasn't the kind of father who said, "That's all right, son, you go play. I'll take over." He was the kind who told me sternly that I'd better learn how to weld that bar back together, then head right back out to the field with it to finish my job. He helped me with the welding and I reattached the bar and carried on. Later that same day, when I broke the bar again in exactly the same way, no one needed to yell at me. I was so frustrated about my own foolishness that I started yelling at myself. Then I asked my father to help me weld the bar back together again and headed out to the fields a third time, quite a bit more cautiously.

Growing up on a farm was great for instilling patience, which was necessary given our rural location. Getting to the enrichment program involved a 2-hour bus ride each way. By the time I was in high school and on the bus only 2 hours a day, total, I felt lucky. On the plus side, I'd long ago got in the habit of using travel time to read and study—I kept trying to do the things an astronaut would do, though it wasn't an exercise in grim obsession. Determined as I was to be ready, just in case I ever got to go to space, I was equally determined to enjoy myself. If my choices had been making me miserable, I couldn't have continued. I lack the gene for martyrdom.

Fortunately, my interests dovetailed perfectly with those of the Apollo-era astronauts. Most were fighter pilots and test pilots;

I also loved airplanes. When I was 13, just as Dave had and my younger brother and sisters would later, I'd joined Air Cadets, which is sort of like a cross between Boy Scouts and the Air Force: you learn about military discipline and leadership, and you're taught how to fly. At 15 I got my glider license, and at 16, I started learning to fly powered planes. I loved the sensation, the speed, the challenge of trying to execute maneuvers with some degree of elegance. I wanted to be a better pilot not only because it fit in with the just-in-case astronaut scenario, but because I loved flying.

Of course, I had other interests, too: reading science fiction, playing guitar, water-skiing. I also skied downhill competitively, and what I loved about racing was the same thing I loved about flying: learning to manage speed and power effectively, so that you can tear along, concentrating on making the next turn or swoop or glide, yet still be enough in control that you don't wipe out. In my late teens I even became an instructor, but although skiing all day was a ridiculously fun way to make money, I knew that spending a few years bumming around on the hills would not help me become an astronaut.

Throughout all this I never felt that I'd be a failure in life if I didn't get to space. Since the odds of becoming an astronaut were nonexistent, I knew it would be pretty silly to hang my sense of self-worth on it. My attitude was more, "It's probably not going to happen, but I should do things that keep me moving in the right direction, just in case—and I should be sure those things interest me, so that whatever happens, I'm happy."

Back then, much more than today, the route to NASA was via the military, so after high school I decided to apply to military college. At the very least, I'd wind up with a good education and an opportunity to serve my country (plus, I'd be paid to go to school). At college I majored in mechanical engineering, thinking that if I didn't make it as a military pilot, maybe I could be an engineer—I'd always liked figuring out how things work. And as I studied and worked numbers, my eyes would sometimes drift up to the picture of the Space Shuttle I'd hung over my desk.

* * *

The Christmas of 1981, six months before graduation, I did something that likely influenced the course of my life more than anything else I've done. I got married. Helene and I had been dating since high school, and she'd already graduated from university and was a rising star at the insurance agency where she worked—so successful that we were able to buy a house in Kitchener, Ontario, before we even got married. During our first two years of wedded bliss, we were apart for almost 18 months. I went to Moose Jaw, Saskatchewan, to begin basic jet training with the Canadian Forces; Helene gave birth to our first child, Kyle, and began raising him alone in Kitchener because a recession had made it impossible to sell our house; we came very close to bankruptcy. Helene gave up her job and she and Kyle moved to Moose Jaw to live in base housing—and then I was posted to Cold Lake, Alberta, to learn to fly fighters, first CF-5s, then CF-18s. It was, in other words, the kind of opening chapter that makes or breaks a marriage, and the stress didn't decrease when, in 1983, the Canadian government recruited and selected its first six astronauts. My dream finally seemed marginally more possible. From that point onward, I was even more motivated to focus on my career; one reason our marriage has flourished is that Helene enthusiastically endorses the concept of going all out in the pursuit of a goal.

A lot of people who meet us remark that it can't be easy being married to a highly driven, take-charge overachiever who views moving house as a sport, and I have to confess that it is not—being married to Helene has at times been difficult for me. She's intimidatingly capable. Parachute her into any city in the world and within 24 hours she'll have lined up an apartment, furnished it with IKEA stuff she gaily assembled herself and scored tickets to the sold-out concert. She raised our three children, often functioning as a single parent because of the amount of time I was on the road, while holding down a variety of demanding jobs, from running the SAP system of a large company to working as a professional chef. She is an über-doer, exactly the kind of person you want riding shotgun when you're chasing a big goal and also trying to have a life. While achieving both things may not take a village, it sure does take a team.

This became extremely clear to me when I was finishing my training to fly fighters and was told I'd be posted to Germany. Helene was very pregnant with our second child, and we were excited about the prospect of moving to Europe. We were already mentally vacationing in Paris with our beautifully behaved, trilingual children when word came down that there had been a change of plans. We were going to Bagotville, Quebec, where I'd fly CF-18s for the North American Aerospace Defense Command (NORAD), intercepting Soviet aircraft that strayed into Canadian airspace. It was a great opportunity to be posted to a brand-new squadron, and Bagotville has much to recommend it, but it is very cold in the winter and it is not Europe in any season. The next three years were difficult for our family. We were still reeling financially, I was flying fighters (not a low-stress occupation) and Helene was at home with two rambunctious little boys—Evan was born just days before we moved to Bagotville—and no real career prospects. Then, when Evan was 7 months old, she discovered she was pregnant again. At the time, it felt to both of us less like a happy accident than the last straw. I looked around, trying to picture what life would be like for us at 45, and thought it would be really hard if I continued to fly fighters. The squadron commanders were working their tails off for not much more money than I was already making; the workload was enormous, there was very little recognition and there was nothing even vaguely cushy about the job. Aside from anything else, being a fighter pilot is dangerous. We were losing at least one close friend every year.

So when I heard Air Canada was hiring, I decided it was time to be realistic. Working for an airline would be an easier life for us, one whose rhythms I already knew well. I actually went to an initial class to get my civilian pilot ratings and then Helene intervened. She said, "You don't really want to be an airline pilot. You wouldn't be happy and then I wouldn't be happy. Don't give up on being an astronaut—I can't let you do that to yourself or to us. Let's wait just a little bit longer and see how things play out."

So I stayed on the squadron and eventually got a tiny taste of being a test pilot: when an airplane came out of maintenance, I would do the test flight. I was hooked. Fighter pilots live to fly, but while I love flying, I lived to understand airplanes: why they do certain things, how to make them perform even better. People on the squadron were genuinely puzzled when I said I wanted to go to test pilot school. Why would anyone give up the glory of being a fighter pilot to be an engineer, essentially? But the engineering aspects of the job were exactly what appealed to me, along with the opportunity to make high-performance aircraft safer.

Canada doesn't have its own test pilot school, but usually sends two pilots a year to study in France, the U.K. or the U.S. In 1987, I won the lottery: I was selected to go to the French school, which is on the Mediterranean. We rented the perfect house there, which came complete with a car. We packed our things, we had goodbye parties. And then, two weeks before we were to wrangle our three kids onto the plane—Kristin was about 9 months old—there was some sort of high-level dispute between the Canadian and French governments. France gave my slot away to a pilot from another country. To say it was a big disappointment personally and a major setback professionally is to understate the case. We were beside ourselves. We'd hit a dead end.

* * *

As I have discovered again and again, things are never as bad (or as good) as they seem at the time. In retrospect, the heartbreaking disaster may be revealed as a lucky twist of fate, and so it was with losing the French slot in the spring. A few months later, I was selected to go to the U.S. Air Force Test Pilot School (TPS) at Edwards Air Force Base, and our year there changed everything. It started out perfectly: we headed to sunny Southern California in December, just as winter gripped Bagotville. Unfortunately, we couldn't go into base housing until the moving van arrived with our furniture. Fortunately, that took several weeks, and in the meantime, we got to spend Christmas at a hotel in Disneyland.

The next year, 1988, was one of the busiest and best of my life. Test pilot school was like getting a Ph.D. in flying; in a single year we flew 32 different types of planes and were tested every day. It was incredibly tough—and incredibly fun: everyone in the class lived on the same street, and we were all in our late 20s or early 30s and liked to have a good time. The program suited me better than anything I'd done to that point, because of its focus on the analytical aspects of flying, the math, the science—and the camaraderie. It was the first time, really, that I'd been part of a group of people who were so much like me. Most of us wanted to be astronauts, and we didn't need to keep our desire a secret anymore. TPS is a direct pipeline to NASA; two of my classmates, my good friends Susan Helms and Rick Husband, made it and became astronauts.

It wasn't at all clear, though, if test pilot school would be a route to the Canadian Space Agency (CSA). When, or even whether, the CSA would select more astronauts was anyone's guess. Only one thing was certain: the first Canadian astronauts were all payload specialists—scientists, not pilots. By that point, though, I'd already committed to trying to follow the typical American path to becoming an astronaut. Maybe I'd wind up with the wrong stuff for the only space agency where I had the right passport, but it was too late to change tack. On the plus side, however, even if I never became an astronaut, I knew I'd feel I was doing something worthwhile with my life if I spent the rest of it as a test pilot.

Our class toured the Johnson Space Center in Houston and visited other flight test centers, like the one in Cold Lake, Alberta, and the Patuxent River Naval Air Station in Maryland, where I ran into a Canadian test pilot who was there as part of a regular exchange program. This guy casually mentioned that his tour was going to end soon and he'd be heading back to Cold Lake, so he guessed someone would be sent to replace him but he wasn't sure who, yet. When I told Helene about this later, she gave me an are-you-thinking-what-I'm-thinking look.

I was. Pax is one of the few major test centers in the world. They have the resources to do cutting-edge work such as testing new types of engines and new configurations for military aircraft, not just for the U.S. but for many other countries, from Australia to Kuwait. Not surprisingly, given the relative size of the Canadian military, Cold Lake tests many fewer planes and focuses on modifications, not on expanding the planes' fundamental capabilities. We had loved living in Cold Lake while I was training to fly fighters, but we'd be spending many years there after I finished test pilot school—why not try to get a stint at Pax first? And yes, there was something else, too: we had become accustomed to warm winters. So I called my career manager (a military officer whose job it is to figure out which billets need to be filled and who could best fill them) and said, "Hey, it would save the Forces about $50,000 if, rather than move us all the way back up to Cold Lake and some other family down to Pax River, you just moved us straight out to Maryland." He was unequivocal: "No way. You're coming back." Oh well, it had been worth a try. But the fact of the matter was that the Canadian government had spent about a million dollars to send me to test pilot school. They had every right to tell me where to go.

We started getting ready to move again. But a month later, I got a phone call from the career manager: "I've got a great idea. How about I send you straight to Pax River?" It probably didn't hurt my case that I was the top graduate that year at TPS and had led the team whose research project got top honors. That was a big deal for me, personally, and I took some nationalistic pride in it, too—a Canadian, the top U.S. Air Force test pilot graduate! I was even interviewed by a reporter for the Cold Lake newspaper. No one at the paper could think of a title for the article, though, so they called out to the test center, and whoever answered the phone said, "Just call it 'Canadian Wins Top Test Pilot' or something to that effect." A friend mailed me a copy of the article, which was a nice keepsake as well as a reality check for my ego. The headline that ran? "Canadian Wins Top Test Pilot or Something to that Effect."

Helene and I decided to make a family vacation out of our move to Pax River, so in December 1988, we packed up our light blue station wagon with fake wooden side panels, a hideous looking vehicle we called The Limo, and drove from California to Maryland. We were a young couple with three little kids, seeing the southern states for the very first time: we went to SeaWorld, explored caves, spent December 25 in Baton Rouge—it was a great adventure.

So was our time at Pax. We rented a farmhouse instead of living in base housing, which was a nice change for everyone. After a while Helene got a job as a realtor because the hours were somewhat flexible; Kyle, Evan and Kristin all eventually started school. And I tested F-18s, deliberately putting them out of control way up high, then figuring out how to recover as they fell to Earth. At first I was pretty tentative, because I'd spent my life trying to control airplanes, not send them ripping all over the place, but as I gained confidence I started trying different techniques. By the end I was hooked on the feeling: just how far out of control could I get the plane to go? In that program we developed some good recovery techniques, counterintuitive ones that wound up saving planes as well as pilots' lives.

Meanwhile, I was still thinking about what qualifications I would need if the CSA ever started hiring again. An advanced degree seemed like a must, so I worked evenings and weekends to complete a master's degree in aviation systems at the University of Tennessee, which had a great distance learning program. I only had to show up to defend my thesis. Probably my most significant accomplishment at Pax River, though, was to pilot the first flight test of an external burning hydrogen propulsion engine, an engine that would make a plane fly far faster than the speed of sound. The paper that Sharon Houck, the flight test engineer, and I wrote about our research won The Society of Experimental Test Pilots' top award. For us, it was like winning an Oscar, not least because the ceremony was held in Beverly Hills and the audience included legendary pilots like Scott Crossfield, the first person in the world to fly at Mach 2, twice the speed of sound.

To cap it all off, I was named the U.S. Navy test pilot of the year in 1991. My tour was drawing to a close and I'd achieved the American dream—citizenship notwithstanding. My plan was to relax a bit and enjoy our final year in Maryland, spend more time with the kids and play a little more guitar. And then the Canadian Space Agency took out an ad in the newspaper.

Wanted: Astronauts.

* * *

I had about 10 feverish days to write and submit my resumé. Helene and I set about making this thing the most impressive document ever to emerge from rural Maryland. Certainly it was one of the most voluminous: there were pages and pages, listing everything I'd ever done, every honor and award and course I could remember. This was back in the day of the dot matrix printer, so we decided we should get it professionally printed, on high quality paper. Then Helene decreed it should be bound, too. That would catch their eye! A professionally bound resumé, approximately the size of a phone book. But we didn't stop there: I had a francophone friend translate the entire thing into perfect French, and we had that version separately printed and bound. We proofed both documents so many times that at night I was dreaming about errant commas, and then we seriously debated driving to Ottawa so we could be 100 percent certain my application got there on time. Reluctantly, I agreed to trust a courier—then called the CSA to be sure the package had actually arrived. It had, along with 5,329 other applications. That was January 1992.

What followed was the least comfortable five-month period of my life. I kept trying to do everything right but there was no feedback and no way to tell if I was succeeding or not.

We heard nothing for weeks, but finally a letter arrived: I'd made it to the top 500 round! The next step was to fill out some psychiatric evaluation forms. I did, and the response was, "You'll hear from us, yes or no, within a few weeks." The "few weeks" came and went. Radio silence. Another week dragged by. Had I come off as so psychologically unbalanced that they were concerned to tell me I was a "no"? Eventually I couldn't stand the uncertainty any longer and phoned the CSA. The guy who answered said, "Wait a minute, let me look at the list. Hadfield. Hmmm . . . Oh yeah, here's Hadfield. Congratulations, you've made it to the next level." Not for the last time, I wondered whether this whole process was in fact a cunningly designed stress test to see how applicants coped with uncertainty and irritation.

By this point, there were 100 of us left. I was asked to go to Washington, D.C., for an interview with an industrial psychologist, who met me in the lobby of a hotel and announced, "I didn't rent a hall or anything, we'll just talk in my room." As we headed up there, all I could think was that if I were a woman, I really would not be feeling good about this at all. When we got to his room, he invited me to make myself comfortable, and I hesitated: bed or chair—which would say the right thing about me? I opted for the chair and answered some questions that were fairly obviously intended to reveal little more than severe psychoses. If I remember correctly, he asked whether I'd ever wanted to kill my mother.

More weeks of waiting, but the phone did finally ring: 50 of us had been given the nod to go to Toronto for more interviews.

Fifty! At this point I did allow myself to believe I had a chance of being selected, and decided it was time to tell my career manager what I was up to. In the U.S., the military pre-selects applicants; you apply to your service and they decide whose names to put forward to NASA. But in Canada, the military had no role in the process, and I think they were rather confused when I called and said, "Thought I should let you know that I've applied to be an astronaut, so you might need to replace me at Pax River a little earlier than planned—or not."

Nothing was much clearer to me after Toronto, where I had initial medical tests to make sure I was basically healthy, as well as a lengthy panel interview with a few CSA people, including Bob Thirsk, one of the first Canadian astronauts. I went back to Maryland, where Helene was excited and confident, and I tried to lead my normal life but could not forget for a moment what was hanging in the balance. For so long, becoming an astronaut had been a theoretical concept, but now that it was really happening—or not—it was horribly nerve-wracking. Would the 9-year-old boy achieve his dreams?

Then I made the final round. Twenty candidates were being summoned to Ottawa at the end of April for a week, so they could get a really good look at us. I was already exercising and eating carefully, but now I really got serious. I wanted to be sure my cholesterol was low—I knew they'd put us under the microscope, medically speaking—and that I was the picture of good health. I figured out the 100 things they might ask me and practiced my answers. Then I practiced them in French. When I got to Ottawa my first thought was that I had some serious competition. The other 19 applicants were impressive. Some had Ph.D.s. Some were military college graduates like me. Some had reams of publications to their names. There were doctors and scientists and test pilots, and everyone was trying to project casual magnificence. Of course, the set-up could not have been more anxiety inducing. No one even knew how many of us might make the final cut. Six? One? I was trying to appear serenely unconcerned while subtly implying that I was the obvious choice, with all the qualifications they were looking for. I hoped.

It was a busy week. There was a mock press conference, to see whether we were skilled at public relations or could be trained to do it. There were in-depth medical exams involving many vials of bodily fluids and a great deal of poking and prodding. But the real make-or-break event was an hour-long panel interview, which included CSA bigwigs, PR people and astronauts. I thought about it all week: How to stand out, yet not be a jerk? What were the best answers to the obvious questions? What should I not say? I'm pretty sure I was the last interview of the week, but in any event the panel members were clearly accustomed to one another's interviewing styles and in the habit of deferring to Mac Evans, who later went on to head the CSA. When it was time to answer a question, they'd say, "Mac, you want to take this one?" I felt I'd bonded a little with these people over the past week, and when someone asked me a really tough question, it just popped out of my mouth: "Mac, you want to take this one?" It was a gamble and could have come off as arrogance, but they laughed uproariously, which bought me another minute to think up a decent answer. However, there was no actual feedback. I had no idea whether they liked me more or less than anyone else. I headed back to Maryland having no clue whether they were going to choose me or not.

In parting, we'd been told that on a particular Saturday in May, all 20 of us would get a phone call between 1:00 and 3:00 p.m. to confirm whether we'd been selected or rejected. When that

Saturday finally arrived, I decided the best thing to do to make the time pass more quickly would be to go water-skiing with friends who had a boat, so that's what we did. Then Helene and I went back to the house to eat lunch and watch the clock. We figured they'd call the people they wanted to hire first, so if someone declined, they could move on to the next name on the list. We were right: shortly after 1:00 the phone rang, and I picked it up in the kitchen. It was Mac Evans, asking if I wanted to be an astronaut.

I did, of course. I always had.

But my main emotion was not joy or surprise or even huge enthusiasm. It was an enormous rush of relief, as though a vast internal dam of self-imposed pressure had finally burst. I had not let myself down. I had not let Helene down. I had not let my family down. This thing we'd worked toward all this time was actually going to happen. Mac told me I could tell my family, as long as they understood it needed to be kept entirely under wraps, so after Helene and I absorbed the news—insofar as we could—I called my mother and swore her to secrecy. She must have started phoning people as soon as she hung up. By the time I got my grandfather on the line, it was old news.

In the subsequent months, there would be excitement, a secret meeting with the other three new astronauts, then hoopla and publicity, even some pomp and circumstance. But the day I got the call from the CSA, I felt as though I'd suddenly, safely, reached the summit of a mountain I'd been climbing since I was 9 years old, and was now looking over the other side. It was impossible, yet it had happened. I was an astronaut.

Only, as it turned out, I wasn't yet. Becoming an astronaut, someone who reliably makes good decisions when the consequences really matter, takes more than a phone call. It's not something anyone else can confer on you, actually. It takes years of serious, sustained effort, because you need to build a new knowledge base, develop your physical capabilities and dramatically expand your technical skill set. But the most important thing you need to change? Your mind. You need to learn to think like an astronaut.

I was just getting started.

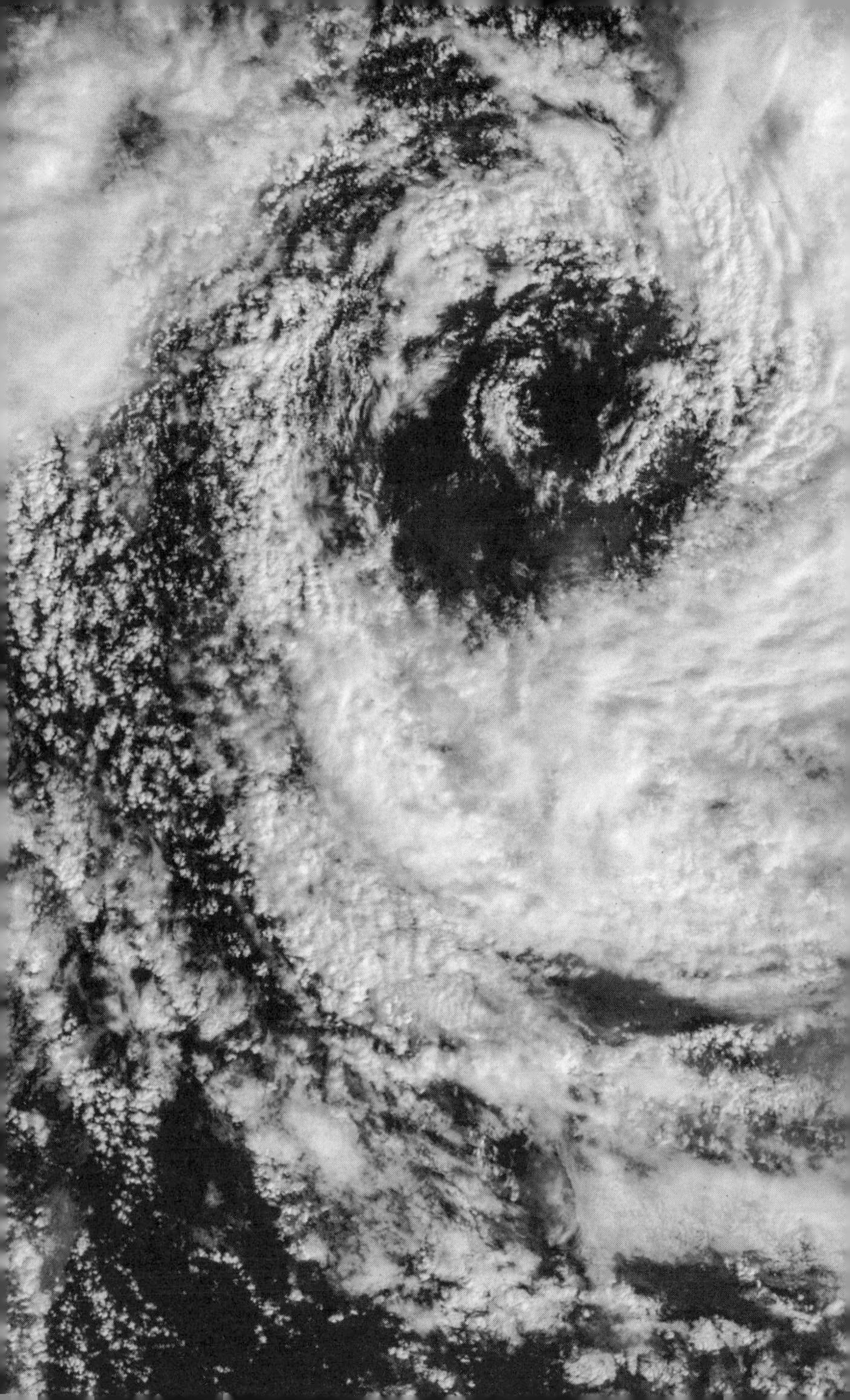

PART I

PRE-LAUNCH

1
THE TRIP TAKES A LIFETIME

One morning a strange thought occurs to me shortly after waking: the socks I am about to put on are the ones I'll wear to leave Earth. That prospect feels real yet surreal, the way a particularly vivid dream does. The feeling intensifies at breakfast, when reporters jostle each other to get a good photo, as though I'm a condemned man and this is my last meal. Similarly, a little later on, when the technicians help me into my custom-made spacesuit for pressure checks, the joviality feels forced. It's the moment of truth. The suit needs to function perfectly—it is what will keep me alive and able to breathe if the spacecraft depressurizes in the vacuum of space—because this isn't a run-through.

I am actually leaving the planet today.

Or not, I remind myself. There are still hours to go, hours when anything could go wrong and the launch could be scrubbed. That thought, combined with the fact that I'm now wearing a diaper just in case we get stuck on the launch pad for a very long time, steers my interior monologue away from the portentous and toward the practical. There's a lot to remember. Focus.

Once everyone in the crew is suited up, we all get into the elevator in crew quarters to ride down to the ground and out to our rocket ship. It's one of those space-age moments I dreamed about as a little kid, except for the slow—*really* slow—elevator. Descent from the third floor takes only slightly less time than it does to boil an egg. When we finally head outside to walk toward the big silver Astro van that will take us to the launch pad, it's that moment everyone knows: flashbulbs pop in the pre-dawn darkness, the crowd cheers, we wave and smile. In the van, we can see the rocket in the distance, lit up and shining, an obelisk. In reality, of course, it's a 4.5-megaton bomb loaded with explosive fuel, which is why everyone else is driving away from it.

At the launch pad, we ride the elevator up—this one moves at a good clip—and one by one we crawl into the vehicle on our hands and knees. Then the closeout crew helps strap me tightly into my tiny seat, and one of them hands me a note from Helene, telling me she loves me. I'm not exactly comfortable—the spacesuit is bulky and hot, the cabin is cramped, a distinctly un-cushion-like parachute and survival kit is wedged awkwardly behind my back—and I'm going to be stuck in this position for a few hours, minimum. But I can't imagine any place else I'd rather be.

After the ground crew checks the cockpit one last time, says goodbye and closes the hatch, it's time for pressure checks of the cabin. Banter ebbs: everyone is hyper-focused. This is all about increasing our chances of staying alive. Yet there's still a whiff of make-believe to the exercise because any number of things could still happen—a fault in the wiring, a problem with a fuel tank—to downgrade this to just another elaborate dress rehearsal.

But as every second passes, the odds improve that we're going to space today. As we work through huge checklists—reviewing and clearing all caution and warning alarms, making sure the multiple frequencies used to communicate with Launch Control and Mission Control are all functional—the vehicle rumbles to life: systems power up, the engine bells chime for launch. When the auxiliary power units fire up, the rocket's vibration becomes more insistent. In my earpiece, I hear the final checks from the key console positions, and my crewmates' breathing, then a heartfelt farewell from the Launch Director. I go through my checklist a quick hundred times or so to make sure I remember all the critical things that are about to happen, what my role will be and what I'll do if things start going wrong.

And now there are just 30 seconds left and the rocket stirs like a living thing with a will of its own and I permit myself to move past hoping to knowing: we are going to lift off. Even if we have to abort the mission after a few minutes in the air, leaving this launch pad is a sure thing.

Six seconds to go. The engines start to light, and we sway forward as this huge new force bends the vehicle, which lurches sideways then twangs back to vertical. And at that moment there's an enormous, violent vibration and rattle. It feels as though we're being shaken in a huge dog's jaws, then seized by its giant, unseen master and hurled straight up into the sky, away from Earth. It feels like magic, like winning, like a dream.

It also feels as though a huge truck going at top speed just smashed into the side of us. Perfectly normal, apparently, and we'd been warned to expect it. So I just keep "hawking it," flipping through my tables and checklists and staring at the buttons and lights over my head, scanning the computers for signs of trouble, trying not to blink. The launch tower is long gone and we're roaring upward, pinned down increasingly emphatically in our seats as the vehicle burns fuel, gets lighter and, 45 seconds later, pushes past the speed of sound. Thirty seconds after that, we're flying higher and faster than the Concorde ever did: Mach 2 and still revving up. It's like being in a dragster, just flooring it. Two minutes after liftoff we're hurtling along at six times the speed of sound when the solid rocket boosters explode off the vehicle and we surge forward again. I'm still completely focused on my checklist, but out of the corner of my eye, I register that the color of the sky has gone from light blue to dark blue to black.

And then, suddenly, calm: we reach Mach 25, orbital speed, the engines wind down, and I notice little motes of dust floating lazily upward. Upward. Experimentally, I let go of my checklist for a few seconds and watch it hover, then drift off serenely, instead of thumping to the ground. I feel like a little kid, like a sorcerer, like the luckiest person alive. I am in space, weightless, and getting here only took 8 minutes and 42 seconds.

Give or take a few thousand days of training.

* * *

That was my first launch, on Space Shuttle *Atlantis*, years ago now: November 12, 1995. But the experience still feels so vivid and immediate that it seems inaccurate, somehow, to describe it in the past tense. Launch is overwhelming on a sensory level: all that speed and all that power, then abruptly, the violence of momentum gives way to the gentle dreaminess of floating on an invisible cushion of air.

I don't think it would be possible to grow accustomed to such an intense experience or be blasé about it. On that first mission, the most seasoned astronaut on board was Jerry Ross, a frequent flyer on the Shuttle. It was his fifth space flight (he subsequently flew twice more, and is one of only two astronauts who've ever launched to space seven times, the other being Franklin Ramón Chang Díaz). Jerry is quietly competent and immensely calm and controlled, the embodiment of the trustworthy, loyal, courteous and brave astronaut archetype. Throughout our training, whenever I was unsure what to do I'd look over to see what he was doing. On *Atlantis*, five minutes before liftoff I noticed he was doing something I'd never seen him do before: his right knee was bouncing up and down slightly. I remember thinking, "Wow, something really incredible must be about to happen if Jerry's knee is bouncing!"

I doubt he was conscious of his own physical reactions. I sure wasn't. I was far too focused on the novelty of what was going on around me to be looking inward. In fact, during ascent, I was checking tables, doing my job, tracking everything I was supposed to track when I suddenly became aware that my face hurt. Then I realized: I'd been smiling so much, without even being aware of it, that my cheeks were cramping up.

More than a quarter-century after I'd stood in a clearing on Stag Island and gazed up at the night sky, I was finally up there myself, orbiting Earth as a mission specialist on STS-74. Our main objective: to construct a docking module on the Russian space station Mir. The plan was use the Shuttle's robot arm to move a newly built docking module up out of its nest in *Atlantis*'s payload bay; install the module on top of the Shuttle; then rendezvous and dock it and *Atlantis* with the station so that future Shuttle flights would have a safer, easier way to get on board Mir than we did.

It was an enormously complicated challenge and we had no way of knowing whether the plan would even work. No one had ever tried to do such a thing before. As it happened, our eight-day mission didn't come off without a hitch. In fact, key equipment failed at a critical moment and nothing proceeded exactly as planned. Yet we managed to construct that docking module anyway, and leaving the station I felt—the whole crew felt—a sense of satisfaction bordering on jubilation. We'd done something difficult and done it well. Mission accomplished. Dream realized.

Only, it hadn't been, not fully anyway. In one sense I felt at peace: I'd been to space at last and it had been even more fulfilling than I'd imagined. But I hadn't been given a lot of responsibility up there—no one is on the first flight—nor had I contributed as much as I would have liked. The difference between Jerry Ross and me, in terms of what we could contribute, was huge. Training in Houston, I hadn't been able to separate out the vital from the trivial, to differentiate between what was going to keep me alive in an emergency and what was esoteric and interesting but not crucial. There had been so much to learn, I'd just been trying to cram it all into my brain. During the mission, too, I was in receive mode: *tell me everything, keep teaching me, I'm going to soak up every last drop*.

So despite having traveled 3.4 million miles, I didn't feel I'd arrived at my destination. An astronaut was something I was still in the process of becoming.

Space flight alone doesn't do the trick. These days, anyone who has deep enough pockets and good enough health can go to space. Space flight participants, commonly known as space tourists, pay between $20 and $40 million each to leave Earth for 10 days or so and go to the International Space Station (ISS) via Soyuz, the compact Russian rocket that is now the only way for humans to get to the ISS. It's not as simple as getting on a plane; they have to complete about six months of basic safety training. But being a space flight participant is not really the same as being an astronaut.

An astronaut is someone who's able to make good decisions quickly, with incomplete information, when the consequences really matter. I didn't miraculously become one either, after just eight days in space. But I did get in touch with the fact that I didn't even know what I didn't know. I still had a lot to learn, and I'd have to learn it the same place everyone learns to be an astronaut: right here on Earth.

* * *

Sometimes when people find out I'm an astronaut, they ask, "So what do you *do* when you're not flying in space?" They have the impression that between launches, we pretty much sit around in a waiting room in Houston trying to catch our breath before the next liftoff. Since you usually only hear about astronauts when they're in space, or about to be, this is not an unreasonable assumption. I always feel I'm disappointing people when I tell them the truth: we are earthbound, training, most of our working lives.

Fundamentally, astronauts are in the service profession: we're public servants, government employees who are tasked with doing something difficult on behalf of the people of our country. It's a responsibility we can't help but take seriously; millions of dollars are invested in our training, and we're entrusted with equipment that's worth billions. The job description is not to experience yee-haw personal thrills in space, but to help make space exploration safer and more scientifically productive—not for ourselves but for others. So although we learn the key skills we will need to know if we go to space, like spacewalking, we spend a lot of our time troubleshooting for other astronauts, helping to work through technical problems that colleagues are experiencing on orbit and also trying to develop new tools and procedures to be used in the future. Most days, we train and take classes—lots of them—and exams. In the evenings and on weekends, we study. On top of that we have ground jobs, supporting other astronauts' missions, and these are crucially important for developing our own skills, too.

Over the years I've had a lot of different roles, from sitting on committees to serving as Chief of International Space Station Operations in Houston. The ground job I held the longest and where I felt I contributed the most, though, was CAPCOM, or capsule communicator. The CAPCOM is the main conduit of information between Mission Control and astronauts on orbit, and the job is an endless challenge, like a crossword puzzle that expands as fast as you can fill it in.

Mission Control Center (MCC) at the Johnson Space Center (JSC) has got to be one of the most formidable and intellectually stimulating classrooms in the world. Everyone in the room has hard-won expertise in a particular technical area, and they are like spiders, exquisitely sensitive to any vibration in their webs, ready to pounce on problems and efficiently dispose of them. The CAPCOM never has anything close to the same depth of technical knowledge but, rather, is the voice of operational reason. I started in 1996 and quickly discovered that having flown even once gave me insight into what it made sense to ask a crew to do in space, and equally important, when. If one of the experts at Mission Control suggested the crew do X, I would be aware of some of the logistical difficulties that someone who'd never been up there might not consider; similarly, the crew knew I could empathize with and understand their needs and challenges because I'd been to space myself. The CAPCOM is less a middleman, though, than an interpreter who is constantly analyzing all changing inputs and factors, making countless quick small judgments and decisions, then passing them on to the crew and the ground team in Houston. It's like being coach, quarterback, water boy and cheerleader, all in one.

Within about a year, I was Chief CAPCOM, and in total worked 25 Shuttle flights. The job had only one drawback: when a launch was delayed, as they often were at Cape Canaveral because of the weather, it could wreak havoc with family vacation plans. Sadly, CAPCOMs cannot telecommute. Other than that, however, I viewed it as a plum assignment, one learning opportunity after another. I learned how to summarize and distill the acronym-charged, technical discussions that were going on over the internal voice loops in Mission Control in order to relay the essential information to the crew with clarity and, I hoped, good humor. When not on console at JSC, I trained with crews to see firsthand how the astronauts interacted and what their individual strengths and weaknesses were, which helped ensure that I could advocate effectively for them when they were in space—and also that I stayed up-to-date in terms of both training and using complex equipment and hardware. I loved the job, not least because I could feel, see and remember my direct contribution to every mission. After each landing, as that crew's plaque was hung on the wall at MCC, I could look up and see not just a colorful symbol of collective accomplishment, but a personal symbol of challenges overcome, complexity mastered, the near-impossible achieved.

When I went to space again on STS-100 in April 2001, it was with a much deeper understanding of the whole puzzle of space flight, not just my own small piece of it. I'm not going to pretend that I wouldn't have welcomed the chance to go to space earlier (American astronauts were, understandably, at the front of the line for Shuttle assignments—the vehicle was made in the U.S.A. and owned by the U.S. government). But without question, being on the ground for six years between my first and second flights made me a much better astronaut and one who had more to contribute both on Earth and off it.

I began training for STS-100 a full four years before we were scheduled to blast off. Our destination, the International Space Station, did not even exist yet; the first pieces of the Station were sent up in 1998. Our main objective was to take up and install Canadarm2, a huge, external robotic arm for capturing satellites and spaceships, moving supplies and people around and, most important, assembling the rest of the ISS. The Shuttle would continue to bring up modules and labs, and Canadarm2 would help place them where they were supposed to go. It was the world's most expensive and sophisticated construction tool, and getting it up and working would require not one EVA (extra vehicular activity, or spacewalk) but two—and I was EV1, lead spacewalker, though I'd never been outside a spaceship in my life.

Spacewalking is like rock climbing, weightlifting, repairing a small engine and performing an intricate pas de deux—simultaneously, while encased in a bulky suit that's scraping your knuckles, fingertips and collarbone raw. In zero gravity, many easy tasks become incredibly difficult. Just turning a wrench to loosen a bolt can be like trying to change a tire while wearing ice skates and goalie mitts. Each spacewalk, therefore, is a highly choreographed multi-year effort involving hundreds of people and a lot of unrecognized, dogged work to ensure that all the details—and all the contingencies—have been thought through. Hyper-planning is necessary because any EVA is dangerous. You're venturing out into a vacuum that is entirely hostile to life. If you get into trouble, you can't just hightail it back inside the spaceship.

I practiced spacewalking in the Neutral Buoyancy Lab, which is essentially a giant pool at JSC, for years. Literally. My experience both during my first flight and at Mission Control had taught me how to prioritize better, how to figure out what was actually important as opposed to just nice-to-know. The key things to understand were what the outside of the ISS would be like, how to move around out there without damaging anything and how to make repairs and adjustments in real time. My goal in the pool was to practice each step and action I would take until it became second nature.

I'm glad I did that, because I ran into some unanticipated problems during the spacewalk, ones I probably couldn't have worked through if my preparation had been slapdash. Ultimately, STS-100 was a complete success: we returned home on Space Shuttle *Endeavour* tired but proud of what we'd accomplished. Helping to install Canadarm2 and playing a part in building this permanent human habitat off our planet—which is all the more remarkable because it has required the participation and cooperation of 15 nations—made me feel like a contributing, competent astronaut.

That feeling didn't diminish even slightly when I proceeded to spend the next 11 years on Earth. I hoped to go back to space, yes, but I wasn't sitting around in space explorers' purgatory, doing nothing. In Star City, where Yuri Gagarin trained, I worked as NASA's Director of Operations in Russia from 2001 to 2003, and I learned to live the local life, really embrace it, in order to understand the people I worked with and be more effective in the role. That experience came in handy when, a decade later, I wound up living and working closely with Russian cosmonauts. Not only did I speak their language, but I knew something about myself: it takes me longer to understand when the culture is not my own, so I have to consciously resist the urge to hurry things along and push my own expectations on others.

From Star City I moved back to Houston to become Chief of Robotics for the NASA Astronaut Office during one of the lowest points in NASA's history. It was 2003, right after the *Columbia* disaster; the Shuttle was grounded, construction on the ISS had therefore ceased, and many Americans were grimly questioning why tax dollars were being spent on such a dangerous endeavor as space exploration in the first place. It seemed possible that while we might overcome the technical hurdles and make the Shuttle a much safer vehicle, we might not be able to roll back the tide of public opinion. Yet we managed to do both, a good reminder of how important it is to retain a strong sense of purpose and optimism even when a goal seems impossible to achieve.

Impossible was, frankly, what a third space flight was starting to look like for me. But just as I had back in college, I decided it made sense to be as ready as I could be, just in case. And so from 2006 to 2008, I was Chief of International Space Station Operations in the NASA Astronaut Office, responsible for everything to do with selection, training, certification, support, recovery, rehab and reintegration of all ISS crew members. Interacting with space agencies in other countries and focusing so intensively on the ISS turned out to be good preparation. I got the nod for another mission: this time, a long-duration expedition.

On December 19, 2012, I went back to space for the third time, via the Russian Soyuz, along with NASA astronaut Tom Marshburn and Russian cosmonaut Roman Romanenko. Crews on the ISS overlap so newcomers have a few months to learn from old-timers; we joined Expedition 34, which was commanded by Kevin Ford. When his crew left in early March 2013, Expedition 35 began with a new commander: me. It was what I'd been working toward my whole life, really, to be capable and competent to assume responsibility for both the crew—which numbered six again in late March, when another Soyuz arrived—and the ISS itself. It was reality, yet hard to believe.

* * *

As I got ready for my third flight, it struck me: I was one of the most senior astronauts in the office. This was not my favorite revelation of all time, given that I didn't—still don't—think of myself as that old. On the plus side, however, people listened to what I had to say and respected my opinion; I had influence over the training and flight design process and could help make it more practical and relevant. Twenty years after I got that phone call from Mac Evans, asking if I wanted to join the CSA, I was an éminence grise at JSC—I'd only been in space 20 days, yet I had turned myself into an astronaut. Or to be more accurate, I'd been turned into an astronaut; NASA and the CSA had seen to that, by providing the right education and experiences.

That third mission, of course, greatly expanded my experience. I didn't just visit space: I got to *live* there. By the time our crew landed, after 146 days in space, we'd orbited Earth 2,336 times and traveled almost 62 million miles. We'd also completed a record amount of science on the ISS. Expedition 34/35 was the pinnacle of my career, and the culmination of years of training—not just training to develop specific job-related skills, like piloting a Soyuz, but training to develop new instincts, new ways of thinking, new habits. And that journey, even more than the ones I've taken in rocket ships, transformed me in ways I could not have imagined when I was a 9-year-old boy looking up at the night sky, transfixed by wonder.

See, a funny thing happened on the way to space: I learned how to live better and more happily here on Earth. Over time, I learned how to anticipate problems in order to prevent them, and how to respond effectively in critical situations. I learned how to neutralize fear, how to stay focused and how to succeed.

And many of the techniques I learned were fairly simple though counterintuitive—crisp inversions of snappy aphorisms, in some cases. Astronauts are taught that the best way to reduce stress is to sweat the small stuff. We're trained to look on the dark side and to imagine the worst things that could possibly happen. In fact, in simulators, one of the most common questions we learn to ask ourselves is, "Okay, what's the next thing that will kill me?" We also learn that acting like an astronaut means helping one another's families at launch—by taking their food orders, running their errands, holding their purses and dashing out to buy diapers. Of course, much of what we learn is technically complex, but some of it is surprisingly down-to-earth. Every astronaut can fix a busted toilet—we have to do it all the time in space—and we all know how to pack meticulously, the way we have to in the Soyuz, where every last item must be strapped down just so or the weight and balance get thrown off.

The upshot of all this is that we become competent, which is the most important quality to have if you're an astronaut—or, frankly, anyone, anywhere, who is striving to succeed at anything at all. Competence means keeping your head in a crisis, sticking with a task even when it seems hopeless, and improvising good solutions to tough problems when every second counts. It encompasses ingenuity, determination and being prepared for anything.

Astronauts have these qualities not because we're smarter than everyone else (though let's face it, you do need a certain amount of intellectual horsepower to be able to fix a toilet). It's because we are taught to view the world—and ourselves—differently. My shorthand for it is "thinking like an astronaut." But you don't have to go to space to learn to do that.

It's mostly a matter of changing your perspective.

2 HAVE AN ATTITUDE

No matter how competent or how seasoned, every astronaut is essentially a perpetual student, forever cramming for the next test. It's not how I envisioned things when I was 9 years old. Then I dreamed of blasting off in a blaze of glory to explore the universe, not sitting in a classroom studying orbital mechanics. In Russian. But as it happens, I love my job—the day-to-day reality of it, not just the flying around in space part (though that is definitely cool).

If the only thing you really enjoyed was whipping around Earth in a spaceship, you'd hate being an astronaut. The ratio of prep time to time on orbit is *many months: single day in space*. You train for a few years, minimum, before you're even assigned to a space mission; training for a specific mission then takes between two and four years, and is much more intensive and rigorous than general training. You practice tricky, repetitive tasks as well as highly challenging ones to the point of exhaustion, and you're away from home more than half the time. If you don't love the job, that time will not fly. Nor will the months after a flight, when you're recovering, undergoing medical testing and debriefing on all kinds of technical and scientific details. Nor will the years of regular training between missions, when you're recertifying and learning new skills, while helping other astronauts get ready for their flights. If you viewed training as a dreary chore, not only would you be unhappy every day, but your sense of self-worth and professional purpose would be shattered if you were scrubbed from a mission—or never got one.

Some astronauts never do. They train, they do all the work and they never leave Earth. I took this job knowing that I might be one of them.

I'm a realist, and one who grew up in a time when "Canadian astronauts" simply didn't exist. I was already an adult, with a university degree and a job, when Canada selected its first astronauts in 1983. So when I finally did get to Houston in 1992, I was elated that it was possible for me to be there at all, but also skeptical about my prospects of leaving the planet. Crew time on the ISS was determined by the amount of money a country contributed; Canada provided less than 2 percent of the Station's funding, so got less than 2 percent of the crew time—an entirely fair and inflexible arrangement. But even Americans who are selected for the astronaut corps have no guarantee that they'll get to space. There's always the possibility of a radical shift in government funding; when programs are canceled, it affects a whole generation of astronauts. Or a rocket might blow up and kill a crew, and then human space flight would be put on hold for years, until a full accident review could be carried out and the public could be convinced it was safe, and worthwhile, to resume. Or the vehicles themselves could change. The Shuttle was retired in 2011, after 30 years in service, and today the Soyuz, a much smaller vehicle, is the only way for human beings to get to the ISS. Some astronauts hired during the Shuttle era are simply too tall to fly in the tiny Soyuz. The possibility that they'll leave Earth is currently zero.

Changes in your own life also affect your chances of flying. You could develop a minor health problem that nevertheless disqualifies you (you have to pass the toughest medical in the world to get to the International Space Station—no one wants to cut a mission short and spend millions of dollars, literally, to bring an ailing astronaut home early). Or a major family crisis could force you to miss your one window of opportunity.

Over time, even the qualifications required to get assigned to a mission can change. The Shuttle carried a crew of seven who were in space just a couple of weeks, so there was room for people whose skill set was deep but not wide. If 12 tons of equipment were being transported to the ISS and everything had to be painstakingly unloaded, reassembled and installed, and then the cargo bay needed to be repacked with a huge load of assorted bits and pieces to take back to Earth, being a fanatically organized loadmaster was qualification enough. On the Soyuz, there's simply not room to fly someone whose main contribution is expertise in a single area. The Russian rocket ship only carries three people, and between them they need to cover off a huge matrix of skills. Some are obvious: piloting the rocket, spacewalking, operating the robotic elements of the ISS like Canadarm2, being able to repair things that break on Station, conducting and monitoring the numerous scientific experiments on board. But since the crew is going to be away from civilization for many months, they also need to be able to do things like perform basic surgery and dentistry, program a computer and rewire an electrical panel, take professional-quality photographs and conduct a press conference—and get along harmoniously with colleagues, 24/7, in a confined space.

In the Shuttle era, NASA wanted people who could operate the most complicated vehicle in the world for short stints. Today,

NASA looks for people who can be locked in a tin can for six months and excel, so temperament alone could disqualify you for space flight. A certain personality type that was perfectly acceptable, even stereotypical, in the past—the real hard-ass, say—is not wanted on the voyage when it is going to be a long one.

* * *

Getting to space depends on many variables and circumstances that are entirely beyond an individual astronaut's control, so it always made sense to me to view space flight as a bonus, not an entitlement. And like any bonus, it would be foolhardy to bank on it. Fortunately, there's plenty to keep astronauts engaged and enthusiastic about the job. I relished the physicality of working in simulators and in the pool, while others thrived on carrying out scientific research and still others liked having input into space policy and helping run the program. Sure, we occasionally grumbled about rules and requirements we didn't like, but "take this job and shove it" are not words you're ever going to hear coming out of an astronaut's mouth. I've never met anyone who doesn't feel it's a job full of dreams.

Taking the attitude that I might never get to space—and then, after I did get there, that I might never go back—helped me hold onto that feeling for more than two decades. Because I didn't hang everything—my sense of self-worth, my happiness, my professional identity—on space flight, I was excited to go to work every single day, even during the 11 years after my second mission when I didn't fly and was, at one point, told definitively that I never would again (more on that later).

It sounds strange, probably, but having a pessimistic view of my own prospects helped me love my job. I'd argue it even had a positive effect on my career: because I love learning new things, I volunteered for a lot of extra classes, which bulked up my qualifications, which in turn increased my opportunities at NASA. However, success, to me, never was and still isn't about lifting off in a rocket (though that sure felt like a great achievement). Success is feeling good about the work you do throughout the long, unheralded journey that may or may not wind up at the launch pad. You can't view training solely as a stepping stone to something loftier. It's got to be an end in itself.

The secret is to try to enjoy it. I never viewed training as some onerous duty I had to carry out while praying fervently for another space mission. For me, the appeal was similar to that of a *New York Times* crossword puzzle: training is hard and fun and stretches my mind, so I feel good when I persevere and finish—and I also feel ready to do it all over again.

In space flight, "attitude" refers to orientation: which direction your vehicle is pointing relative to the sun, Earth and other spacecraft. If you lose control of your attitude, two things happen: the vehicle starts to tumble and spin, disorienting everyone on board, and it also strays from its course, which, if you're short on time or fuel, could mean the difference between life and death. In the Soyuz, for example, we use every cue from every available source—periscope, multiple sensors, the horizon—to monitor our attitude constantly and adjust if necessary. We never want to lose attitude, since maintaining attitude is fundamental to success.

In my experience, something similar is true on Earth. Ultimately, I don't determine whether I arrive at the desired professional destination. Too many variables are out of my control. There's really just one thing I can control: my attitude during the journey, which is what keeps me feeling steady and stable, and what keeps me headed in the right direction. So I consciously monitor and correct, if necessary, because losing attitude would be far worse than not achieving my goal.

* * *

My kids are endlessly amused by what they see as my earnestness. For years now they have played a game they call "The Colonel Says," which involves parroting sayings of mine that they find particularly hilarious. My son Evan's personal favorite, which I barked at him from beneath the family car I was trying to fix: "No one ever accomplished anything great sitting down." Recently, they've joked about creating a "Colonel Says" app that would spit out sayings appropriate to any situation. It's a great idea, though I think you'd only need one: "Be ready. Work. Hard. Enjoy it!" It fits every situation.

Think about *Survivor*, which Helene and I have been known to watch on occasion. The show has been on for years now, so everybody knows some of the skills you need in order to win: how to make a fire, for instance, and build a shelter out of branches. And yet, year after year, contestants show up without knowing the basics. I don't get that. You knew you were going to be on *Survivor*—were you just counting on good looks and charm to catch a fish? Knowing that the stakes are a million dollars and a whole different life, why not come prepared?

To me, it's simple: if you've got the time, use it to get ready. What else could you possibly have to do that's more important? Yes, maybe you'll learn how to do a few things you'll never wind up actually needing to do, but that's a much better problem to have than needing to do something and having no clue where to start.

This isn't just how I approach my job. It's how I live my life. For instance, a few years ago I was invited to take part in an air show in Windsor, Ontario, that was scheduled to overlap with an Elton John concert. The organizers decided to try to get him to cross-promote the air show. I thought the chances of a superstar interrupting his performance to promote a regional air show were quite slim, but then I started wondering: What if he agreed? What if it turned out that Elton John was a fanatic about airplanes or, secretly, a space geek—what was the most extreme thing that might wind up happening?

I've played the guitar since I was a kid. While I'm not the best guitarist in the world, I do love it, and for years I've played and sung in bands on Earth, including the all-astronaut band Max Q, and in space, too. A vision, not an entirely pleasant one, flashed before my eyes: Elton John somehow finding this out and inviting the guitar-playing astronaut from the air show up on stage to strum a few bars with him. The likelihood of that was almost zero, I knew that, but I'd performed with the Houston Symphony, so I also knew that unlikely things do occur sometimes. So my next thought was, "All right, let's say that did happen—what song would he ask me to play?" There was only one possible answer: "Rocket Man." So I sat down and learned how to play it and practiced to the point where I was reasonably confident I wouldn't be booed off the stage. I actually started kind of hoping I would get to go up and play "Rocket Man" with Elton John.

As it happened, I did wind up at the concert, and Helene and I did get to meet Elton John and we had a very nice, normal 10-minute conversation with him. But I never got anywhere near the stage nor, to this moment, is Elton John aware that I can pull off a respectable rendition of his song. But I don't regret being ready.

That's how I approach just about everything. I spend my life getting ready to play "Rocket Man." I picture the most demanding challenge; I visualize what I would need to know how to do to meet it; then I practice until I reach a level of competence where I'm comfortable that I'll be able to perform. It's what I've always done, ever since I decided I wanted to be an astronaut in 1969, and that conscious, methodical approach to preparation is the main reason I got to Houston. I never stopped getting ready. Just in case.

If, when I was 21, someone had asked me to write a film script for the life I wanted, it would've gone like this: fighter pilot, test pilot, astronaut. Happy marriage, healthy kids, interesting experiences. My life has followed that script, but there were so many "ifs" that could have changed the plot: if, for instance, I hadn't seen the CSA's newspaper ad soliciting applications—which could well have happened, since we were living in the U.S. at the time. However, I never thought, "If I don't make it as an astronaut, I'm a failure." The script would have changed a lot if, instead, I'd moved up in the military or become a university professor or a commercial test pilot, but the result wouldn't have been a horror movie.

* * *

I didn't walk into JSC a good astronaut. No one does. The most you can hope for is that you're good astronaut material. Some people who make it through the selection process turn out not to be, and what makes the difference is that quality I mentioned earlier: attitude. You have to be willing to sit in Russian classes for years, and willing to train repetitively on safety procedures on board the ISS even though you think you know them inside out. You have to accept that you'll need to master a lot of skills that seem arcane, or that you might never even get to use, or both. And you can't view any of it as a waste of time.

Even better is if you can view it all as being fun or at least interesting. In 2001, I became Director of Operations in Russia for NASA, a job that, back then, was not coveted by most American astronauts. Historic tensions between the two countries were off-putting to some, while others simply weren't thrilled about the idea of being immersed in a foreign culture, complete with a different alphabet, brutal winters and a dearth of modern conveniences such as dishwashers and clothes dryers. To a Canadian who'd managed to acclimatize to the drawl and humidity of Gulf Coast Texas, however, the chance to live in yet another foreign country for a few years sounded exciting, so I was happy to get the posting. Wanting to make the most of our time there, I took extra Russian classes, as did Helene (our three kids were all at boarding school or university in Canada); she telecommuted to her job in Houston so she could spend most of each month with me in Star City, about an hour outside of Moscow, which is where cosmonauts train. Instead of moving into one of the American townhouses that NASA built there, we decided to live in a Russian apartment building, figuring that would improve our chances of really getting to know the country and people.

And it did. We were forced to speak the language, and we had great evening get-togethers with our neighbors that featured music, dancing and communal *shashlik*, the delicious Russian version of barbecue. Memorably, one of NASA's local drivers, Valodya, decided to initiate me into the semi-mystical process of selecting, cutting and preparing the meat for *shashlik*, which takes half a day, followed by just two days to recover. There was vodka to bless the meat, Moldovan cognac to toast the genealogy of the swine, Russian beer to sip while cutting cubes of semi-frozen pork, red wine to marinate the mixture and yourself, and, as the day went on, increasingly emotional speeches about the beauty of raw meat and the bond of kinship between men. Valodya and I chopped up 170 pounds of meat as well as whole bags of onions and tomatoes, then mixed in dusty pouches of herbs and spices as we drank every bottle of liquid in his home, all while watching grainy soccer on a 10-inch TV. By the end of the evening there were five great teeming buckets of fermenting pork to be thrown on the fire the next day, we were closer than family (a good thing, as I left my coat, hat, camera and keys at Valodya's place) and I took great pride in not throwing up in the van that came to take me home. Best of all, the time-honored recipe we so carefully followed remains a complete secret, as I can't really remember exactly what we did.

However, it would be disingenuous to pretend that I viewed the job in Russia solely as an entertaining foreign adventure. The Shuttle was already slated for retirement and the Soyuz would, by the end of the decade, be the main mode of transportation to the ISS. Clearly, the partnership between the U.S. and Russia was going to become increasingly important. Learning the language and figuring out how Roscosmos, the Russian space agency, operates was all part of getting ready for the big changes everyone knew were coming, and being sure that I was still qualified to fly. Just in case.

It's never either-or, never enjoyment versus advancement, so long as you conceive of advancement in terms of learning rather than climbing to the next rung of the professional ladder. You *are* getting ahead if you learn, even if you wind up staying on the same rung. That's why I asked if I could be trained to fly the Soyuz. I was interested in the vehicle itself—it's so different from the Shuttle—though I knew my chances of actually getting to fly it were about the same as my chances of jamming on stage with Elton John. A North American would have to be in space with a completely incapacitated Russian commander in order to ever be allowed to fly the thing. And before that, you'd have to be assigned to a mission. A long line of dominoes would have to fall in a very unusual way, in other words.

I thought that maybe it would pay off one day—but if not, hey, flying a Soyuz was an interesting thing to know how to do and maybe I'd pick up skills that would transfer to some other area. So I got qualified to be a flight engineer cosmonaut and to perform spacewalks in the Russian spacesuit. That extra training ate into my free time, obviously. But it also wound up giving me insight into the Russian system, which is significantly different from ours in terms of its greater emphasis on academic mastery before you ever start simulating. Understanding their perspective wound up helping me in my day job, especially when I was trying to negotiate conflicts between our space program and theirs. I've never been called on to command the Soyuz nor spacewalk for Russia, and I never will be. But I'm still glad I know how.

Some astronaut training is very much like going to school: you sit in a classroom with an instructor, get tested and receive grades. But we also train on computers and in simulators that are full-scale mock-ups of actual spacecraft. At JSC, my favorite place to train is in the pool. Sometimes we're in the Neutral Buoyancy Lab to develop hardware and test new procedures for future missions. Sometimes we're trying to work out solutions to problems faced by astronauts who are currently on orbit; on Earth, where the stakes are low, we have a lot more latitude to experiment. But we also do a lot of training in the lab because floating in water is as close as we can get on Earth to floating in microgravity and it allows us to practice EVAs. I really feel like a full-fledged astronaut in the pool: I'm wearing a spacesuit, my breathing is assisted just as it is during a spacewalk—it's realistically evocative. It is also physically exhausting, but I never tire of it—I spent about 50 full days practicing in the pool before my first spacewalk in 2001. After six hours in the water, I have no trouble falling asleep at night.

A surprising amount of my training has been esoteric, once-in-a-lifetime kind of stuff it would be hard not to love. In the summer of 2010, for instance, I did some work with the international research team at Pavilion Lake in British Columbia. It's a beautiful, clear freshwater lake, the bottom of which is studded with microbialites: rock structures of all different shapes and sizes that look a lot like coral. Microbialites were very common for about two billion years of Earth's early history but are quite rare today. So the purpose of the Pavilion Lake Research Project is to try to figure out how they are forming in order to understand more about the origins of life on Earth. It's kind of like exploring another planet, being down there at the bottom of the lake looking at these things, so the international research team decided it made sense to get astronauts involved. As a result, I got qualified as a DeepWorker pilot. The DeepWorker is an amazing little one-person vehicle, a bit like a personal submarine, that is so fun to operate that some (wealthy) people buy them as toys. You drive with your feet—one pedal moves you vertically, the other horizontally—and manipulate the vehicle's robotic arm with your hands. It's otherworldly, being in your own little waterproof bubble 200 feet underwater, filming and gathering samples of structures that are directly linked to the beginning of life on Earth.

This kind of work is a natural fit for astronauts. We're trained to operate vehicles that require hand, eye and foot coordination in a hostile environment, without slamming into anything. And NASA and the CSA are interested in the project because the study of microbialites may provide tools that will help us identify ancient forms of life on other planets—and because the DeepWorker is an analogue for the kinds of vehicles we may use someday to collect samples on the Moon, an asteroid or Mars. The astronauts who wind up doing that work will need to know how to be the on-the-ground hands and eyes for scientists back on Earth who are counting on them to gather the right information and samples. So the goal is to learn lessons at Pavilion Lake about how to train astronauts to be geologists—not great geologists, just good enough ones—because that makes a lot more sense than trying to train leading geologists to be astronauts.

These are long-range goals, obviously. I'm never going to the Moon or Mars. I may not even be alive when someone else does. A lot of our training is like this: we learn how to do things that contribute in a very small way to a much larger mission but do absolutely nothing for our own career prospects. We spend our days studying and simulating experiences we may never actually have. It's all pretend, really, but we are learning. And that, I think, is the point: learning.

My first space flight, to Mir was in 1995. At the time, it was a big deal because I was the first and only Canadian ever to go on board. No one even remembers that mission today, and Mir has long since been deorbited and burned up in the atmosphere. My first flight is irrelevant to everyone but me. I can let that crush me and spend the rest of my life looking back over my shoulder, or I can maintain attitude. Since that choice is mine, I'll keep on getting ready to play "Rocket Man."

Just in case.

3 THE POWER OF NEGATIVE THINKING

"HOW DO YOU DEAL WITH YOUR FEAR?"

It's one of the questions I'm asked most often. When people think about space exploration, they don't just picture Neil Armstrong stepping off the ladder of the Lunar Module and onto the Moon. They also remember the smoke plume etched in the sky after the Space Shuttle *Challenger* exploded shortly after launch, and the startling, fiery bursts of light as *Columbia* disintegrated on re-entry, raining down metal and human remains. These spectacularly violent images of space flight have been engraved on public consciousness as deeply as the joyfully triumphant ones.

Naturally then, when people try to imagine what it feels like to sit in a rocket with the engines roaring and firing, they assume it must be terrifying. And it *would* be terrifying if you were plucked off the street, hustled into a rocket ship and told you were launching in four minutes—and oh, by the way, one wrong move and you'll kill yourself and everybody else. But I'm not terrified, because I've been trained, for years, by multiple teams of experts who have helped me to think through how to handle just about every conceivable situation that could occur between launch and landing. Like all astronauts, I've taken part in so many highly realistic simulations of space flight that when the engines are finally roaring and firing for real, my main emotion is not fear. It's relief.

At last.

In my experience, fear comes from not knowing what to expect and not feeling you have any control over what's about to happen. When you feel helpless, you're far more afraid than you would be if you knew the facts. If you're not sure what to be alarmed about, everything is alarming.

I know exactly how that feels, because I'm afraid of heights. When I stand near the edge of a cliff or look over the railing of a balcony in a high-rise, my stomach starts tumbling, my palms sweat and my legs don't want to move even though the rising panic in my body insists that I get back to safety. Right now. That physical response doesn't bother me, though. I think everyone *should* be afraid of heights. Like fearing pythons and angry bulls, it's a sensible self-preservation instinct. But I recognize it seems incongruous for a pilot/astronaut to be afraid of heights. How can I possibly do my job when just being up high triggers primal fear?

The answer is that I've learned how to push past fear. Growing up on the farm, my brothers and sisters and I used to go out to our barn, where the grain corn was stored, and climb up to the rafters, then jump down into the corn, just to feel the way the dried kernels suddenly rushed up around our feet and legs, like deep, loose, rounded gravel. So long as we landed feet-first and balanced, we would come to a smooth stop. As we gained confidence, we leapt from higher and higher rafters, until we were jumping from two or three stories up, daring each other, daring ourselves. My fear was there always, strongly, but I wasn't immobilized by it. I always managed to make myself jump. I think I was able to do it because of the gradual buildup in terms of height, the progressive sense of confidence rooted in actual experience and the simple fact that practice made me more skilled.

But my fear of heights didn't go away. When I was a teenager, my dad used to take me flying in his biplane. In the summertime it was warm enough to take the canopy off and fly open cockpit, with nothing at all between us and the sky—or the ground, when my dad flew upside-down and did aerobatics. Initially, suspended headfirst, thousands of feet above the ground, restrained from falling only by a seat belt, I was paralyzed by terror. My hands and arms reflexively braced against the sides of the cockpit, as if holding on would hold me in. Every muscle in my body was tensed, vibrating, and there was a rushing feeling, almost like a noise, going up and down the back of my skull.

Yet I didn't fall out of the plane. The seat belt attached in five places and kept me pinioned, rock-solid, in my seat. My eyes told me that nothing was keeping me from plummeting to my death, but with experience, I started to be able to override that sensation with reason: I was actually just fine, I wasn't going to fall out of the plane. Eventually the fear that I *might* faded.

I'm still scared to stand at the edge of a cliff. But in airplanes and spaceships, while I know I'm up high, I'm also sure I can't fall. The wings and structure and engines and speed all succeed in keeping me up, just as the surface of the Earth holds me up when I'm on the ground. Knowledge and experience have made it possible for me to be relatively comfortable with heights, whether I'm flying a biplane or doing a spacewalk or jumping into a mountain of corn. In each case, I fully understand the challenge, the physics, the mechanics, and I know from personal experience that I'm not helpless. I do have some control.

People tend to think astronauts have the courage of a superhero—or maybe the emotional range of a robot. But in order to stay calm in a high-stress, high-stakes situation, all you really need is knowledge. Sure, you might still feel a little nervous or stressed or hyper-alert. But what you won't feel is terrified.

* * *

Feeling ready to do something doesn't mean feeling certain you'll succeed, though of course that's what you're hoping to do. Truly being ready means understanding what could go wrong—and having a plan to deal with it. You could learn to scuba dive in a resort pool, for instance, and go on to have a wonderful first dive in the ocean even if you had no clue how to buddy breathe or what to do if you lost a flipper. But if conditions were less than ideal, you could find yourself in serious danger. In the ocean, things can go wrong in one breath, and the stakes are life or death. That's why in order to get a scuba license you have to do a bunch of practice dives and learn how to deal with a whole set of problems and emergencies so that you're really ready, not just ready in calm seas.

For the same sort of reasons, trainers in the space program specialize in devising bad-news scenarios for us to act out, over and over again, in increasingly elaborate simulations. We practice what we'll do if there's engine trouble, a computer meltdown, an explosion. Being forced to confront the prospect of failure head-on—to study it, dissect it, tease apart all its components and consequences—really works. After a few years of doing that pretty much daily, you've forged the strongest possible armor to defend against fear: hard-won competence.

Our training pushes us to develop a new set of instincts: instead of reacting to danger with a fight-or-flight adrenaline rush, we're trained to respond unemotionally by immediately prioritizing threats and methodically seeking to defuse them. We go from wanting to bolt for the exit to wanting to engage and understand what's going wrong, then fix it.

Early on during my last stay on the ISS, I was jolted to consciousness in the middle of the night: a loud horn was blaring. For a couple of seconds I was in a fog, trying to figure out what that unpleasant noise was. There were four of us in the American segment of the Station then, and like prairie dogs, we all poked our heads up out of our sleep pods at the same time to look at the panel of emergency lights on the wall that tell us whether we should be concerned about depressurization, toxicity or some other potentially fatal disaster. Suddenly all of us were wide awake. That deafening noise was the fire alarm.

A fire is one of the most dangerous things that can happen in a spaceship because there's nowhere to go; also, flames behave less predictably in weightlessness and are harder to extinguish. In my first year as an astronaut, I think my response to hearing that alarm would have been to grab an extinguisher and start fighting for my life, but over the past 21 years that instinct has been trained out of me and another set of responses has been trained in, represented by three words: warn, gather, work. "Working the problem" is NASA-speak for descending one decision tree after another, methodically looking for a solution until you run out of oxygen. We practice the "warn, gather, work" protocol for responding to fire alarms so frequently that it doesn't just become second nature; it actually supplants our natural instincts. So when we heard the alarm on Station, instead of rushing to don masks and arm ourselves with extinguishers, one astronaut calmly got on the intercom to warn that a fire alarm was going off—maybe the Russians couldn't hear it in their module—while another went to the computer to see which smoke detector was going off. No one was moving in a leisurely fashion, but the response was one of focused curiosity, as though we were dealing with an abstract puzzle rather than an imminent threat to our survival. To an observer it might have looked a little bizarre, actually: no agitation, no barked commands, no haste.

The next step is to gather, so we joined the Russians in their part of the Station to start working the problem. How serious was the threat? So far, all the signs were reassuring. We couldn't smell smoke or see flames. Maybe one little wire had melted somewhere, or the detector was responding to dust. We talked to Mission Control in Houston and in Moscow, but as we investigated, checking the module where the detector had been triggered, it seemed more and more likely that we were dealing with a simple malfunction. Finally everyone agreed that it had been a false alarm, and we headed back to our sleep stations. An hour later, when the fire alarm sounded again, we repeated the warn, gather, work protocol just as before. The response was similarly calm, though not perfunctory—possibly something had been slowly smoldering for the past hour. As it turned out, nothing had. The detector was a lemon, that's all. I remember thinking, "That was just like a sim, only better, because now I get to go to sleep."

I doubt anyone's heart rate increased by more than a beat or two while we were dealing with those fire alarms, even during the first minutes when the threat of a raging inferno seemed most real. We felt competent to deal with whatever happened—a sense of confidence that comes directly from solid preparation. Nothing boosts confidence quite like simulating a disaster, engaging with it fully, both physically and intellectually, and realizing you have the ability to work the problem. Each time you manage to do that your comfort zone expands a little, so if you ever face that particular problem in real life, you're able to think clearly.

You never want to get so comfortable when you're training that you think, "Ho hum, here we go again, playing 'astronaut in peril.'" For a sim to work, you really have to buy into it. Fidelity helps: we train to fight fires on the ISS, for instance, in a full-scale simulator that is pumped full of real smoke—so full that, in one sim our crew did in the service module shortly before my last flight, we couldn't see our own feet by the time we managed to get our gas masks on. As commander, I decided, "The smoke is too thick, let's close the hatches and regroup in another module to figure out how to work the problem." This led to a rather spirited debrief afterward with the Russian team running the exercise. I'd responded perfectly by American standards—NASA trains us to close off the burning segment, save the crew, then figure out how to fight the fire—but the Russian philosophy is different. They want us to stand and fight the fire. Their reasoning is that the rescue vehicle, the Soyuz, is docked at one end of that service module. As I explained to the trainers afterward, we would've been delighted to stop and fight, only, the sim was a little *too* realistic. I had to respond the way I would in real life: in a terrible fire, with such thick smoke, I'd opt to go with NASA's procedures and save the crew, not the lab—after all, we'd still have food, water and communications capability even if we lost the service module. A sim, on Earth, is the right place to expose these kinds of philosophical disconnects and resolve them. Next time we did this sim, the Russians compromised: they filled the service module with a level of smoke that we all agreed made it possible and sensible to stand and fight.

The notion that a fire might break out while we were on the ISS was not hypothetical: in 1997, two years after I visited, an oxygen-generating canister did start a fire on Mir. The crew worked the problem, throwing wet towels on the canister until they extinguished the flame; their spacecraft was smoke-filled and they didn't have enough masks left afterward, but everyone survived. That incident reminded everyone that there's a good reason we train for disaster. Space exploration is inherently dangerous. If my focus ever wavers in the classroom or during an eight-hour simulation, I remind myself of one simple fact: space flight might kill me.

To drive that message home, we have what we euphemistically refer to as "contingency sims"—death sims, actually—which force us to think through our own demise in granular detail: not only how we'd die, but what would happen afterward to our families, colleagues and the space program itself. These are table-top sims, primarily for the benefit of management, so they don't occur in an actual simulator but in a boardroom with people participating via speakerphone if necessary. Everyone who in real life would be involved in dealing with an astronaut's death takes part: doctors, space program administrators, media relations people—even the dead astronaut.

A death sim starts with a scenario—"Chris is seriously injured on orbit," say—and over the next few hours, people work through their own roles and responses. Every five to ten minutes whoever is running the exercise tosses what we call a "green card" into the mix: in essence, a new wrinkle. The cards are devised by the training team, whose job it is to conjure up as many realistic twists and turns as possible; no one else in the sim knows in advance what is on the cards, and we respond as though these things are actually happening. One green card might be, "We've just received word from the Station: Chris is dead." Immediately, people start working the problem. Okay, what are we going to do with his corpse? There are no body bags on Station, so should we shove it in a spacesuit and stick it in a locker? But what about the smell? Should we send it back to Earth on a resupply ship and let it burn up with the rest of the garbage on re-entry? Jettison it during a spacewalk and let it float away into space?

While people are discussing how quickly my body would start to decompose and what kind of help my crewmates might need to deal with the trauma, they are hit with another green card: "Someone has just tweeted that there's been an accident on the ISS, and a *New York Times* reporter is calling to find out what's going on." New problems, while the old ones are still being dealt with: How should the PR people respond? Should NASA or the CSA take the lead? When will a statement be issued and what should it say? The green cards start coming faster and faster, posing new problems, just as would happen in real life: Who should tell my parents their son is dead? By phone or in person? Where will they even be—at the farm or at the cottage? Do we need two plans, then, depending on where my mom and dad are?

As is probably clear by now, death sims are not weepy, grief-stricken affairs. They're all about brass tacks. Although family members aren't required to participate, Helene has joined in several times because she has discovered that taking the time to verbalize what you think you would do in the worst-case scenario quickly reveals whether you're really prepared or not. During a contingency sim before Expedition 34/35, for instance, she realized that her plan to trek in the Himalayas while I was in space for five months was wonderful—unless something went seriously wrong during my mission. The green cards in the sim forced us to figure out who would contact our kids if I died (quite possibly a reporter, we realized, if their mother was on a mountaintop) and how quickly Helene could get to Houston to be with them (not very, considering how many connecting flights she'd need to take). We had to think about the minutiae that would become highly relevant if I died on the ISS: cell reception in remote hill towns in Asia, for instance, and how the difference in time zones would affect her ability to get in touch with key decision-makers in Houston. The upshot of all this was that Helene decided to save the Himalayas for another year and hike in Utah instead. In fact, everyone who participated in the sim discovered weaknesses in their own planning and went back to the drawing board on a few items. (Except me, but that's what happens when you're dead.)

Sometimes a sim is a proving ground where you demonstrate how well-rounded your capabilities are, but more often, it's a crucible where you identify gaps in your knowledge and encounter domino effects that simply never occurred to you before. When I first started training with Roman Romanenko, my crewmate on that last mission and the commander of our Soyuz, we did a re-entry sim together in the simulator in Star City. Roman had actually flown in a Soyuz before and I had not, so my main goal was just to help out where I could. At one point, I noticed that the oxygen tank inside our capsule was leaking a little bit. It didn't seem like a big deal. We had multiple tanks and the leak was tiny. We kept concentrating on the complex tasks associated with re-entry, but then it hit me: that tank is leaking into a really small capsule, which means the oxygen level is rising to the point where everything may become flammable, so now we may have to depressurize the cabin to avoid a fire—but if we do, we may not have enough oxygen to get home.

A normal, gradual re-entry was out of the question. It didn't matter if we were anywhere near Kazakhstan. We had to turn that spaceship around and drop to Earth, immediately, or we'd die. But I didn't know the fastest way to turn the Soyuz around and Roman was already knee-deep in another procedure, so we missed the very narrow window when we still had a chance to save ourselves. What had initially seemed like a subtle failure—a tiny leak in an oxygen tank—wound up killing us.

Roman and I hadn't really understood the operational impact of a leaking tank, but we sure did after that sim, and in subsequent training, we came up with a much better response. A sim is an opportunity to practice but frequently it's also a wake-up call: we really *don't* know exactly what we're doing and we'd better figure it out before we're facing this situation in space.

While play-acting grim scenarios day in and day out may sound like a good recipe for clinical depression, it's actually weirdly uplifting. Rehearsing for catastrophe has made me positive that I have the problem-solving skills to deal with tough situations and come out the other side smiling. For me, this has greatly reduced the mental and emotional clutter that unchecked worrying produces, those random thoughts that hijack your brain at three o'clock in the morning. While I very much hoped not to die in space, I didn't live in fear of it, largely because I'd been made to think through the practicalities: how I'd want my family to get the news, for instance, and which astronaut I should recruit to help my wife cut through the red tape at NASA and the CSA. Before my last space flight (as with each of the earlier ones) I reviewed my will, made sure my financial affairs and taxes were in order, and did all the other things you'd do if you knew you were going to die. But that didn't make me feel like I had one foot in the grave. It actually put my mind at ease and reduced my anxiety about what my family's future would look like if something happened to me. Which meant that when the engines lit up at launch, I was able to focus entirely on the task at hand: arriving alive.

* * *

Although simulating a catastrophe does get you accustomed to the idea that it could happen, you're never inured to the point of indifference. I doubt I will ever be able to forget the morning of February 1, 2003. I'd flown back to Houston from Russia the night before, and forgot to turn my phone back on until Helene and I were driving to brunch in the morning. As soon as I did, I saw I had a massive number of messages; she checked her phone, and so did she. We didn't have to listen to them to know something terrible had happened. Our friends on *Columbia* were coming home that day. We turned the car around and drove back to the house with an awful, awful feeling, like all the air had gone out of everything.

I turned on the TV and immediately there it was, a replay of *Columbia's* disintegration in the skies not all that far from our home. My eyes filled with tears even before I'd really processed the information, and Helene crumpled to her knees, weeping. The sudden, irretrievable loss was devastating. We knew all seven astronauts on that Shuttle. We'd shared the same dream. We cared about their spouses and children. The commander of that mission, Rick Husband, was my classmate at test pilot school; we'd sung together and worked on a research project together. Rick had signed on to help out my family at one of my launches, and wound up cheerfully driving to Orlando when my parents got stranded there and bringing them back to Cape Canaveral. Great guy, close friend. I mourned, and still mourn, his death and the deaths of our six other friends on that flight.

I also felt a huge sense of disappointment and responsibility: I was part of a program that had let this happen. When I got to the office an hour or so later, they were already mounting teams to go help pick up the pieces of our colleagues and their spaceship, which had been scattered across the state because of the way the Shuttle broke apart. I helped out at JSC and did what I could for Rick's family. But there wasn't much anyone could do. Highly talented, hard-working, genuinely nice people had been killed doing their jobs, through no fault of their own. It was a terrible, needless waste.

Yet I never considered leaving NASA, nor was it ever a topic of discussion with my family. I hadn't been assigned to another Shuttle flight and didn't think I ever would be, so there was no threat to my own safety. My job was to help others fly safely, and the *Columbia* disaster only strengthened my sense of purpose. We had to persuade the world all over again that the Shuttle was safe to fly and that the work the crew had been doing was vitally important and should be continued. Like most people at NASA, I felt that accomplishing those two things was the best way to honor *Columbia*'s crew, and I'm sure it's what they would have wanted. I've never known an astronaut who doesn't believe that the work we do is far more important than we are as individuals.

I'm extremely proud to have been part of the effort to figure out how to identify, prevent and mitigate risks so the Shuttle could fly again without harming one more person. There were three things we had to do: one, decrease the chances of damage during ascent; two, figure out a better way to recognize, while the Shuttle was still in space, whether there had been any damage; three, come up with ways to repair damage on orbit. Shortly after *Columbia*, I became Chief of Robotics at the NASA Astronaut Office, responsible for developing space robotics techniques and hardware and making sure astronauts and cosmonauts knew how to use them, so I was very involved in helping figure out solutions to the last two challenges. Actually, every single person in our organization got behind the effort, despite the fact that morale was low and public support for the space program was even lower.

We were entirely successful. We changed how we attached and inspected foam; we devised a way to survey the vehicle once it was on orbit (we repurposed some unused Canadarm hardware to build a kind of boom for the Shuttle, then mounted a camera on it so we could survey all the most fragile parts of the spaceship); we figured out how to use a special type of glue during an EVA to fix any damage—and we always had a rescue Shuttle standing by in case the first one got in trouble. The Shuttle became a much safer vehicle and we never lost another crew member. I never had another opportunity to fly on one, but I would've done so in a heartbeat.

The reason is not that I have a death wish. I'm not even a thrill-seeker. Few astronauts are. Strapping yourself on top of what is essentially a large bomb is plenty risky—there's no need to up the ante. I've never been interested in the just-for-the-hell-of-it rush of, say, bungee jumping. If you're an adrenaline junkie, I understand why you'd find that exciting. But I'm not, and I don't.

To me, the only good reason to take a risk is that there's a decent possibility of a reward that outweighs the hazard. Exploring the edge of the universe and pushing the boundaries of human knowledge and capability strike me as pretty significant rewards, so I accept the risks of being an astronaut, but with an abundance of caution: I want to understand them, manage them and reduce them as much as possible.

It's almost comical that astronauts are stereotyped as daredevils and cowboys. As a rule, we're highly methodical and detail-oriented. Our passion isn't for thrills but for the grindstone, and pressing our noses to it. We have to: we're responsible for equipment that has cost taxpayers many millions of dollars, and the best insurance policy we have on our lives is our own dedication to training. Studying, simulating, practicing until responses become automatic—astronauts don't do all this only to fulfill NASA's requirements. Training is something we do to reduce the odds that we'll die. Sometimes, as with *Challenger* and *Columbia*, a vehicle fails and there's absolutely nothing the crew can do. But sometimes there is. Astronauts have survived fires on the launch pad and in space, ballistic landings where the Soyuz has come back through the atmosphere like a rock hurled from space—even a collision that punctured a spacecraft and caused sudden depressurization. In a real crisis like that, a group hug isn't going to save you. Your only hope is knowing exactly what to do and being able to do it calmly and quickly.

My kids used to make fun of me for having more homework than they did and for taking it a lot more seriously, too. But when the risks are real, you can't wing it. The person that homework should matter to most of all is me. Having safety procedures down cold might save my life someday, and would definitely help me avoid making dumb mistakes that actually increased the risks. No matter how bad a situation is, you can always make it worse. Let's say the Soyuz engines start failing going into deorbit burn, so I shut them off, but then can't start them again—well, I just took a big problem and made it huge.

Preparation is not only about managing external risks, but about limiting the likelihood that you'll unwittingly add to them. When you're the author of your own fate, you don't want to write a tragedy. Aside from anything else, the possibility of a sequel is nonexistent.

* * *

A few years ago our band was playing a gig in Houston when a woman came up to the stage and asked, "Do you know 'Proud Mary'? I'll sing it." She carried herself with supreme confidence and even looked a bit like Tina Turner, so we said, "Sure!" She came on the stage, grabbed the mic with authority, we started playing the song—and she didn't start singing. I thought, "Oh, she just doesn't know where to come in," so I helped her with the first verse. But, it quickly became apparent, the only words she actually knew were "Rolling on the river." She'd belt those out at the appropriate moments and then kind of hum her way through the rest of the lyrics. Clearly, she'd assumed that as soon as she had a microphone in her hand, she'd magically turn into Tina Turner. Perhaps even more foolishly, we'd just assumed that she was prepared. That was a big assumption given the North American subculture of pretense, where watching *Top Chef* is the same thing as knowing how to cook.

When the stakes are high, preparation is everything. In my day job, the stakes are highest during dynamic operations, when variables change rapidly, triggering chain reactions that unfold in a hurry. Now, this isn't always the case in space. Sometimes you have a fair amount of time to deal with a problem, even a serious one. The ISS, for instance, drifts around the world like a miniature moon, with no engines firing, and would continue to do so even after a complete electrical failure. Everything could fizzle out, reducing the Station to a lifeless hulk, but we'd be fine for days—enough time to attempt quite a few different repairs and then, if nothing worked, bail out and head back to Earth in our Soyuz. If, however, a small meteorite smacked into the side of the Station—suddenly, you're into dynamic ops. Now there's a timeline, every second counts, and you'd better do things in the correct sequence or you're going to die.

The most dynamic operations occur during launch and deorbit burn, when engines are firing, so we simulate contingencies and malfunctions during those two phases of space flight hundreds if not thousands of times. If the engine malfunctions during deorbit burn in the Soyuz, for instance, you know you're not going to re-enter the atmosphere the way you wanted. Maybe you won't land where rescue vehicles are waiting to meet you. Maybe instead of pulling 4 g, or four times the force of gravity on Earth, it will be more like 8 or 9 g, which is not just extremely uncomfortable but also more dangerous; plus, you'll need extra strength, given the physical pressure on your body, simply to reach up and flip the switches that control the vehicle. Or maybe the rocket won't be set up right and you'll skip off the atmosphere, like a stone across a pond, and then not have enough fuel left to attempt the deorbit burn later. Or maybe the Soyuz will simply break into pieces and burn up in the atmosphere.

Whatever happens, it's going to happen fast, and your survival will to a large extent depend on your competence. The interactions—between the vehicle's own internal systems, its actual velocity and attitude, how far it is from Earth—are really complicated. It *is* rocket science. You have to understand what causes which effects, and you have no time to explain things to your crewmates or to yourself. You really need to know what it means if you're 20 degrees off attitude, or what to do if one of your thrusters fails, as well as the dozens of follow-on consequences that will trigger yet more chain reactions. You don't even have a few seconds to wrack your brain—you need that information right now, front of mind, in order to make a good decision.

In training, once we understand the theory and the basics of the interactions between systems, we start learning what it looks like when systems fail, one at a time. Initially we do this via "part task trainers," or PTTs, which are one-on-one computer simulations run by an instructor who's usually sitting beside us using a separate laptop. For instance, in a PTT on the thermal control system of the Soyuz, I stared at that system's normal display on my computer screen, getting used to what it should look like, and then the instructor failed one of the pumps so I could see what would happen. Next he showed me how it would change if a sensor failed and it appeared as if we had a temperature regulation problem but really the issue was just that the thermometer had gone haywire. I spent a lot of time on PTTs looking at the symptoms of false alarms versus actual system failures: pressure regulation, atmospheric constituent controls, the rendezvous sensing system—the list is long.

Through this process I started to figure out what to pay attention to and what to disregard, which risks were the greatest and which would trigger the most negative consequences, and then I was ready for the actual Soyuz simulator, to see what the whole picture looked like. My instructors in the control room started with individual failures and over time worked up to integrated failures: the thermal regulation system malfunctions and on top of that, the digital control loop on the central computer fails—how does all that fit together? Do these problems compound each other or are they unrelated? Uh-oh, now an engine has failed and we're on backup thrusters. What are our options?

These sims are all about prioritizing risks, understanding how they interrelate and deciding which ones must be dealt with immediately—all of which you need to figure out well before you get to space, where hesitation could be fatal. On Earth, there's the luxury of time. The instructors can even freeze the simulator to make sure you really get it: "You just lost the digital computer—look at how the vehicle is recalculating acceleration and engine cut-off time, how it's going to control attitude for atmospheric entry. Try to think about each step here."

Eventually, I built up to dealing with cascading malfunctions, where the trainers throw in everything including the kitchen sink. It's like writing a final exam in university where you're scribbling down answers as fast as you possibly can, non-stop, for hours. When I got out of a tough integrated sim, I was whipped. I may have looked calm on the outside, but my brain had just had a brutal workout and was now able to handle no challenge greater than locating a bottle of beer and heading for my back porch.

When I graduated to doing a really challenging sim with my crew, we started preparing *for* the preparation, in order to get the most out of it. Before Roman, Tom Marshburn and I simulated deorbit burn together, for example, we talked about how we were going to handle certain problems—"If the digital computer fails at this point, we're going to work it through this way"—and split up our roles and responsibilities. Each of us had his own thing to be hyper-aware of while the dynamic operations were going on, and we planned out our first three or four actions for a variety of different scenarios, so we were all on the same page. I got in the habit of asking during each sim we did together, "Okay, what's the summary of our failures to this point?" Tom would list them and we'd quickly prioritize them and figure out which ones were still immediate threats.

A lot of people talk about expecting the best but preparing for the worst, but I think that's a seductively misleading concept. There's never just one "worst." Almost always there's a whole spectrum of bad possibilities. The only thing that would really qualify as *the* worst would be not having a plan for how to cope.

* * *

Now for the confusing part: take your simulation seriously and engage as fully as you would in real life—but be prepared that the sim itself may be wrong. This happens to us most often with simulators that are used to train not for disasters but for skill development.

In 1992, for instance, when I was a brand-new astronaut, the maiden voyage of Space Shuttle *Endeavour* was scheduled to rescue an Intelsat VI-F3 satellite that hadn't made it to its required high orbit of 23,000 miles above Earth. Its engine wasn't working properly, so this hugely expensive communications satellite had got stuck drifting along in a low orbit, about 300 miles overhead, where it was completely useless. The plan was that a crew would go to space, strap a new motor onto the thing, then release it to ascend to its intended geostationary orbit. But first, since the Canadarm wasn't designed to latch onto an uncooperative satellite, an astronaut would have to do a spacewalk to install a custom-built grapple fixture while riding on the end of the arm. The grapple fixture could then be used to grab the satellite; it would be sort of like sticking a big handle on the side of it.

The plan was table-topped, and then a simulator was built. Of course, without weightlessness the simulator wouldn't be of much use, so they used a NASA facility that is something like a gigantic air hockey table. The astronaut who was going to grab the satellite practiced over and over on this thing with the Canadarm simulator until he'd developed a good technique for attaching the handle to the satellite. However, even on an air hockey table there's a tiny bit of friction, the implications of which were not fully understood until the astronaut was actually in space. In true weightlessness, he just couldn't get enough force to make the grapple bar latch on before the satellite wobbled away again.

This happened repeatedly until everyone in space and on the ground was cursing the sim. The satellite was a large cylinder that looked a bit like a silver grain silo, so big that an astronaut wouldn't be able to stop it with his hands and might actually be ripped right off the end of the Canadarm if he tried. Two astronauts would have the same problem.

What about three astronauts? That might work. Only, three's a crowd in the Shuttle airlock, which was built to hold two astronauts, max. Also, all three would have to be in position to grab simultaneously—was that even physically possible? And even if it were, how could the commander ever maneuver the Shuttle close enough to the satellite for the attempt to happen? The crew in space got a day off while on Earth, astronauts and trainers began working these separate problems in round-the-clock simultaneous sims, both in the full-scale Shuttle simulator, in order to see how close it could get to a satellite, and in the buoyancy lab to solve the three-astronauts-in-the-airlock riddle and also figure out what the trio would do if they actually did manage to grab the satellite. It was a day of feverish invention, culminating in a fully integrated sim that was run a few times until the powers-that-be agreed: "It's worth a shot."

There was a happy ending: the three astronauts did manage to stop the satellite, install the new motor and send it on its way. Mission accomplished. But although the problem was solved via sims, it was also created by a sim. The moral of the story: part of preparing for the worst is keeping in mind that your sim itself may be based on the wrong assumptions, in which case you'll draw the wrong, perfectly polished conclusions.

* * *

It's puzzling to me that so many self-help gurus urge people to visualize victory, and stop there. Some even insist that if you wish for good things long enough and hard enough, you'll get them—and, conversely, that if you focus on the negative, you actually invite bad things to happen. Why make yourself miserable worrying? Why waste time getting ready for disasters that may never happen?

Anticipating problems and figuring out how to solve them is actually the opposite of worrying: it's productive. Likewise, coming up with a plan of action isn't a waste of time if it gives you peace of mind. While it's true that you may wind up being ready for something that never happens, if the stakes are at all high, it's worth it. Think about driving down the highway listening to the radio and enjoying the sunshine, versus scanning the road, noticing the oil truck up ahead and considering what will happen if, just as you pull out to pass, you're cut off by the van that you've noticed has been driving a little erratically in the left lane for the past 10 minutes. Anticipating that problem would be the best way to avoid it.

You don't have to walk around perpetually braced for disaster, convinced the sky is about to fall. But it sure is a good idea to have some kind of plan for dealing with unpleasant possibilities. For me, that's become a reflexive form of mental discipline not just at work but throughout my life. When I get into a really crowded elevator, for instance, I think, "Okay, what are we going to do if we get stuck?" And I start working through what my own role could be, how I could help solve the problem. On a plane, same thing. As I'm buckling my seat belt, I automatically think about what I'll do if there's a crisis.

But I'm not a nervous or pessimistic person. Really. If anything, I'm annoyingly upbeat, at least according to the experts (my family, of course). I tend to expect things will turn out well and they usually do. My optimism and confidence come not from feeling I'm luckier than other mortals, and they sure don't come from visualizing victory. They're the result of a lifetime spent visualizing defeat and figuring out how to prevent it.

Like most astronauts, I'm pretty sure that I can deal with what life throws at me because I've thought about what to do if things go wrong, as well as right. That's the power of negative thinking.

4 SWEAT THE SMALL STUFF

I GRADUATED FROM MILITARY COLLEGE in 1982 with a degree in mechanical engineering and a clear plan: I was going to be a military pilot. Like most of my classmates with similar ambitions I'd been flying small planes for years, and during the summer of 1980 I'd completed the basic flight training course in Portage la Prairie, Manitoba. But to get my wings I had to go to Moose Jaw, Saskatchewan, to learn to fly jets. The Canadian Forces basic jet training course was demanding: 200 hours in a CT-114 Tutor (a two-seater that's now primarily used by the Royal Canadian Air Force aerobatics team, the Snowbirds), accompanied by an instructor who evaluated every flight. If you flew poorly even one time, you were sent for extra training and then had to repeat the flight. Usually, though, that was the beginning of a downward spiral: a "re-ride" was a big black mark on a pilot's record. If you got too many, you were kicked out of the program. It didn't help that each re-ride was posted on a huge board in plain view. If your name appeared there, other pilots started to treat you as if you were already halfway out the door. It was very difficult to recover your confidence, and a lot of trainees simply couldn't do it.

Every flight, then, was make or break. For me, the stakes felt particularly high: it was 1983, the year Canada selected its first astronauts and my impossible dream was becoming ever so slightly less impossible—but only if I flew fighter jets, the traditional first step on the path to becoming an astronaut. There was just one way to be certain I'd get to fly fighters, and that was to ace the jet training course. Only the top graduate got to pick whether to go fighters or transport (flying large planes to move troops and cargo), or be an instructor; no one else in the class had any say in the matter. So I was determined to finish first. The odds of becoming an astronaut were very low, to say the least, but if I didn't become a fighter pilot, they'd be zero.

Then I made a hash out of one of my instrument exam flights. The instructor was someone I'd never flown with before, so he had no idea whether I was any good or not, and I gave him plenty of reasons to think I wasn't. I flew clumsily and didn't prepare properly to transition from one phase of instrument flight to the next; the whole time, I was "behind the airplane," hanging on and reacting rather than anticipating and controlling the vehicle accordingly. The instructor noticed every dumb mistake and criticized me roundly, then grimly started flipping through my record. It was clear he was on the verge of scheduling a re-ride.

Academic failure was new to me—between hard work and natural ability I'd always been successful. It didn't occur to me to try to defend myself, because the guy was right. I'd messed up. I just sat there mute with shame, staring straight ahead and listening to the sound of pages being turned.

After a very long minute, the instructor finally looked up from my file and said, "I see this is the first flight where you've had problems like this, so I'm going to chalk it up to a bad day. No re-ride."

It was not just a reprieve but a life-changing moment. If he hadn't given me the benefit of the doubt, I might well not be an astronaut today. It still haunts me, how close I came to blowing my own chances. Even at the time, the moral of the story was unmistakable: I couldn't afford to be unprepared in any situation where I was going to be evaluated, formally or not. I had to be ready, always.

I decided to change the way I prepared, effective immediately. At night, instead of studying in my room, I studied in the airplane I'd be flying the next day. I got out all the checklists and navigation procedures, and acted out the whole flight, pretending to use the instrument controls. Once I was done and had "landed" safely, I started all over again. No one told me to go sit in a cold hangar for a couple of hours and run through the flight repeatedly until I could picture the whole thing. No one had to. That near re-ride had redoubled my resolve to finish first so I could fly fighters. And it just seemed like common sense that I'd fly the Tutor much better if, when I got in the plane with an instructor the next day, it was (at least mentally) the fourth time I'd made that particular flight.

I also started trying to visualize the route in detail beforehand. "All right, I'm going to go up to Speedy Creek, cut across to Regina—what does that look like, in reality?" When you're 200 feet above the ground, going 240 knots, you want to know where you are at all times, but it's easy to get lost on the prairie. From the air, a lot of southern Saskatchewan looks a lot like the rest of southern Saskatchewan: vast, flat, green and brown treeless fields, bordered by the grid lines of roads and occasionally punctuated by a dry lake bed or the jagged scar of a valley. On my days off, I got in the habit of driving out to where I'd be flying that week and getting out of the car to take a good look around. It paid off. Many times I'd be flying along and suddenly recognize something: "Hey, that's where I parked, I remember that road—I know exactly where I am."

This wasn't just a beginner's tactic, by the way. Even after accumulating thousands of flying hours in high-performance aircraft, I still did something similar. For a complicated flight in an F-18, for instance, I'd get a map of the region and draw my route on it, though I knew I'd never actually see the ground once I was airborne; I figured out which navigational aids I'd be able to use and what that meant for switch throws in the cockpit; I reviewed my checklists, just as I had the very first time I flew a fighter. The point of all this was so that when I was up in the air and actually flying, it already felt familiar. (Plus, I just like to understand exactly where I am—especially on the International Space Station, where I appreciate the view of a sprawling city nestled on a river between quiet volcanoes even more when I know I am looking at Taipei, Taiwan.)

When you think about it, this sort of intensive preparation and play-acting is a permissible form of cheating. It's a bit like telling your opponent in the middle of a game of chess, "Hey, I want to take a break with the board just like this, I'll be back in a few hours," then running off and using that time to try dozens of gambits and figure out the three best moves you can make. That extra effort would give you a significant competitive advantage, particularly if the other guy decided to use the time to take a nap.

I viewed jet training as an ongoing test, and my goal was to create every possible advantage for myself and give the best possible answer to every single question. So when I blew that flight and nearly got a re-ride, I was forced to look inside myself to try to figure out why I hadn't been ready. Was I tired? Hungover? Not assertive enough at the controls? Too focused on the wrong things?

No. The problem was simple: I'd decided I was already a pretty good pilot, good enough that I didn't need to fret over every last detail. And it's true, you don't need to obsess over details if you're willing to roll the dice and accept whatever happens. But if you're striving for excellence—whether it's in playing the guitar or flying a jet—there's no such thing as over-preparation. It's your best chance of improving your odds.

In my next line of work, it wasn't even optional. An astronaut who doesn't sweat the small stuff is a dead astronaut.

* * *

In any field, it's a plus if you view criticism as potentially helpful advice rather than as a personal attack. But for an astronaut, depersonalizing criticism is a basic survival skill. If you bristled every time you heard something negative—or stubbornly tuned out the feedback—you'd be toast.

At NASA, everyone's a critic. Over the years, hundreds of people weigh in on our performance on a regular basis. Our biggest blunders are put under the microscope so even more people can be made aware of them: "Check out what Hadfield did—let's be sure no one ever does *that* again."

Often, we're scrutinized and evaluated in real time. Quite a few simulations involve a crowd: all the people in Mission Control who would in real life work that particular problem, plus the trainers who dreamed up the scenario in the first place and the experts who best understand the intricate components of whatever system is being tested. When we're simulating deorbit to landing, for instance, dozens of people observe, hoping that something new—a flaw in a standard procedure, say, or a better way of doing something—will be revealed. They actually *want* us to stumble into a gray zone no one had recognized could be problematic in order to see whether we can figure out what to do. If not, well, it's much better to discover that gray zone while we're still on Earth, where we have the luxury of being able to simulate a bunch more times until we do figure it out. Whether we fail or succeed in a sim is only part of the story. The main point is to learn—and then to review the experience afterward from every possible angle.

The debrief is a cultural staple at NASA, which makes this place a nightmare for people who aren't fond of meetings. During a sim, the flight director or lead astronaut makes notes on major events, and afterward, kicks off the debrief by reviewing the highlights: what went well, what new things were learned, what was already known but needs to be re-emphasized. Then it's a free-for-all. Everyone else dives right in, system by system, to dissect what went wrong or was handled poorly. All the people who are involved in the sim have a chance to comment on how things looked from their consoles, so if you blundered in some way, dozens of people may flag it and enumerate all the negative effects of your actions. It's not a public flogging: the goal is to build up collective wisdom. So the response to an error is never, "No big deal, don't beat yourself up about it." It's "Let's pull on that"—the idea being that a mistake is like a loose thread you should tug on, hard, to see if the whole fabric unravels.

Occasionally the criticism is personal, though, and even when it's constructive, it can sting. Prior to my last mission, my American crewmate Tom Marshburn and I were in the pool for a six-hour EVA evaluation, practicing spacewalking in front of a group of senior trainers and senior astronauts. Tom and I have both done EVAs in space and I thought we did really well in the pool. But in the debrief, after I'd explained my rationale for tethering my body in a particular way so I'd be stable enough to perform a repair, one of our instructors announced to the room, "When Chris talks, he has a very clear and authoritative manner—but don't let yourself be lulled into a feeling of complete confidence that he's right. Yes, he used to be a spacewalking instructor and evaluator and he's Mr. EVA, but he hasn't done a walk since 2001. There have been a lot of changes since then. I don't want the junior trainers to ignore that little voice inside and not question something just because it's being said with authority by someone who's been here a long time."

At first that struck me as a little insulting, because the message boiled down to this: "Mr. EVA" *sounds* like he knows what he's doing, but really, he may not have a clue. Then I stopped to ask myself, "Why is the instructor saying that?" Pretty quickly I had to concede that the point was valid. I don't come off as wishy-washy and I'm used to teaching others how to do things, so I can sound very sure of myself. That doesn't mean I think I know everything there is to know; I'd always assumed that people understood that perfectly well and felt free to jump in and question my judgment. But maybe my demeanor was making that difficult. I decided to test that proposition: instead of waiting for feedback, I'd invite it and see what happened. After a sim, I began asking my trainers and crewmates, "How did I fall short, technically, and what changes could I make next time?" Not surprisingly, the answer was rarely, "Don't change a thing, Chris—everything you do is perfect!" So the debrief did what it was supposed to: it alerted me to a subtle but important issue I was able to address in a way that ultimately improved our crew's chances of success.

At NASA, we're not just expected to respond positively to criticism, but to go one step further and draw attention to our own missteps and miscalculations. It's not easy for hyper-competitive people to talk openly about screw-ups that make them look foolish or incompetent. Management has to create a climate where owning up to mistakes is permissible and colleagues have to agree, collectively, to cut each other some slack.

I got used to public confessionals as a fighter pilot. Every Monday morning we got together for a flight safety briefing and talked about all the things that could have killed us the previous week. Sometimes pilots confessed to really basic errors and oversights and the rest of us were expected to suspend judgment. (Deliberate acts of idiocy—flying under a bridge, say, or showing off by going supersonic over your friend's house and busting every window in the neighborhood—were a different story. Fighter pilots could be and were fired for them.) It was easier not to pass judgment once I grasped that another pilot's willingness to admit he'd made a boneheaded move, and then talk about what had happened next, could save my life. Literally.

At NASA, where the organizational culture focuses so explicitly on education, not just achievement, it's even easier to frame individual mistakes as teachable moments rather than career-ending blunders. I remember one astronaut, also a former test pilot, standing up at a meeting and walking us all through an incident where his T-38 (the plane we all train on to keep up our flying skills) slid off the end of a runway in Louisiana. For a pilot this is hugely embarrassing, a rookie error. There wasn't much damage to the plane, so the guy could've either kept his mouth shut, or the moral of the story could have been, "All's well that ends well." But as he told it, the moral was: be careful because the asphalt at this runway is slicker than most—it contains ground-up seashells, which, it turns out, are seriously slippery when it's raining. That was incredibly useful information for all of us to have. While no one thought more of that astronaut for sliding off the runway, we certainly didn't think less of him for being willing to save us from doing the same thing ourselves.

* * *

After a four-hour sim, we usually debrief for about an hour, but that's nothing. After a space flight, we debrief all day, every day, for a month or more, one subject at a time. Communication systems, biological research, spacesuits—every aspect of each experience is picked apart in an exhaustive meeting with the people responsible for that particular area. We gather in the main conference room of the Astronaut Office at JSC, a windowless, rather cavernous place, and the senior experts in that day's subject matter take seats around a large oval table beside the recently returned astronauts, while the not-so-senior experts sit in chairs lined up against the walls. The flavor of the meetings is grilled astronaut: the experts fire questions at us and we do our best to answer them fully, with as many details as possible. In the debrief about food, for instance, we're asked, "How was it? What did you like? Why? Was there enough for everyone? What did you throw away? How about the packaging—any way you can think of to improve it?" (The level of detail we go into helps explain why the food on Station is, for the most part, really good.)

When the topic of discussion is an unexpected occurrence, such as the unplanned EVA to locate an external ammonia leak on the ISS during my last mission, the debrief goes on for days. As I'll explain later, that was a highly unusual spacewalk for a variety of reasons, and the novelty factor made the debrief especially long and involved. The room was packed with people trying to deconstruct and reconstruct events, and figure out what they could do better next time.

And as in any debrief, everyone also wanted to review what *we* could have done better—and to magnify and advertise our errors, so other astronauts wouldn't make the same ones. One of the main purposes of a debrief is to learn every lesson possible, then fold them back into what we call *Flight Rules* so that everyone in the organization benefits.

Flight Rules are the hard-earned body of knowledge recorded in manuals that list, step by step, what to do if X occurs, and why. Essentially, they are extremely detailed, scenario-specific standard operating procedures. If while I was on board the ISS a cooling system had failed, *Flight Rules* would have provided a blow-by-blow explanation of how to fix the system as well as the rationale behind each step of the procedure.

NASA has been capturing our missteps, disasters and solutions since the early 1960s, when Mercury-era ground teams first started gathering "lessons learned" into a compendium that now lists thousands of problematic situations, from engine failure to busted hatch handles to computer glitches, and their solutions. Our flight procedures are based on these rules, but *Flight Rules* are really for Mission Control, so that when we have problems on orbit they can walk us through what to do.

Given the obsession with preparation, it's interesting how frequently we do run into trouble in space. Despite all our practice runs on Earth, it often turns out that we have miscalculated or overlooked something obvious, and need a new flight rule to cover it. In 2003, when I was Chief of Robotics at NASA, a crew on the ISS came very close to inadvertently hitting a fragile part of a docked Shuttle with Canadarm2. In the debrief afterward, it became obvious that although the impending near-collision had been detected on the ground, there wasn't a clear and simple way to alert the crew. The chain of communication was incredibly convoluted: video and data from orbit were transmitted to Houston, where a specialist in a backroom had to recognize the problem and alert the robotics flight controller in Mission Control, who then had to warn the flight director and the CAPCOM, who then had to understand the situation and tell the astronauts what to do, who then had to do the right thing—and all this had to happen while the robot arm continued moving closer and closer to smashing into the only vehicle that could get the crew home alive.

In the debrief we also realized that although astronauts had been very well prepared to use the relatively simple arm on the Shuttle, which had good lighting in the payload bay and fewer things to hit, they were less well trained to manipulate a more sophisticated robotic arm on a structure as complex and poorly lit as the ISS. So in the calm aftermath, we decided that along with making some changes to training, we'd better come up with a fast and unambiguous response people could use when a problem was observed in real time. Sounds like a no-brainer, right? But none of this had occurred to anyone before. And we had to take into account possibly fuzzy and intermittent radio communications, crew members whose first language might not be English, the actual controls on the robot arm itself and the urgency of the problem that had been detected. What we came up with was the simplest possible radio call and the simplest possible crew reaction: whoever saw that Canadarm2 was getting perilously close to smashing into something would say "all stop" three times. Everyone who heard the command, whether on the ground or in space, would repeat it out loud. And the crew would halt the arm's motion with a single switch. This was captured in a new flight rule, so crews and Mission Control now train with the All-Stop Protocol in mind, and brief it aloud before every robotic operation, both in sims and on orbit. And the robot arm has never hit a structure accidentally.

As is probably clear by now, even making seemingly simple decisions can be extremely difficult in space. The beauty of *Flight Rules* is that they create certainty when we have to make tough calls. For instance, in 1997 I was CAPCOM for STS-83, which, shortly after launch, appeared to have a fuel cell issue. Fuel cells generate electricity, sort of like a battery, and one of the three on board appeared to surpass permissible voltage thresholds. At Mission Control we thought the problem was probably with the sensor, not the fuel cell itself, so we were inclined to ignore it. But *Flight Rules* insisted the fuel cell had to be shut down—and then, with only two fuel cells deemed fully operational, another flight rule kicked in: the mission had to be terminated.

If it had been up to us, STS-83 probably could've kept on going, because the Shuttle would fly fine with just two fuel cells if no other problems cropped up. In real time, the temptation to take a chance is always higher. However, the flight rules were unequivocal: the Shuttle had to return to Earth. As CAPCOM, it was my job to tell the commander, "Listen, I know you just got up there, but you have to come on back. Starting now." It was heartbreaking for the crew, after spending so long training for that specific mission, to return to Earth three days after launch with most of their objectives unfulfilled. I'm sure they cursed the flight rules as they deorbited—and cursed even more loudly later, when it turned out the fuel cell in question would likely have been completely fine if they'd stayed in space. (There's a nice coda to this story: the same crew launched again just three months later—which was unprecedented—and that time, nothing went wrong.)

One reason we're able to keep pushing the boundaries of human capability yet keep people safe is that *Flight Rules* protect against the temptation to take risks, which is strongest when momentum has been building to meet a launch date. The Soyuz can launch in just about any weather but the Shuttle was a much less rugged vehicle, so there were ironclad launch criteria: how windy it could be, how cold, how much cloud cover—clearly spelled-out minimally acceptable weather conditions for a safe launch. We came up with them when there was no urgency or pressure and there was enough time to pull on every string and analyze every consequence. We had to invoke them for about one-third of all launches. Having hard and fast rules, and being unwilling to bend them, was a godsend on launch day, when there was always a temptation to say, "Sure, it's a touch colder than we'd like, but . . . let's just try anyway."

I had helped with so many launches at the Cape that I fully expected a weather delay when I got strapped into my seat on *Atlantis* in November 1995, all ready for my first trip to space. Sure enough, five minutes before we were supposed to launch, STS-74 was called off. The weather was actually beautiful in Florida that day, but it was bad at all of our overseas emergency landing sites. The chances that we'd have to abort the mission after liftoff were extremely slim, but the rules were clear: we needed to have the option. No one on board was delighted with this turn of events, but there wasn't a lot of grousing. After so many years of training, what was one more day? That's one good thing about habitually sweating the small stuff: you learn to be very, very patient. (And we did, in fact, launch the next day.)

NASA's fanaticism about details and rules may seem ridiculously finicky to outsiders. But when astronauts are killed on the job, the reason is almost always an overlooked detail that seemed unimportant at the time. Initially, for instance, astronauts didn't wear pressure suits during launch and re-entry—the idea had been considered but dismissed. Why bother, since they were in a proven vehicle with multiple levels of redundancy? It seemed over-the-top, and besides, suits would take up room, add weight to the rocket and, because they're unwieldy, make it more difficult for the crew to maneuver. The Russians began wearing pressure suits for launch and landing only after a ventilation valve came loose and a Soyuz depressurized during re-entry in 1971, killing all three cosmonauts on board, likely within seconds. Shuttle astronauts started wearing pressure suits only after *Challenger* exploded during launch in 1986. In the case of both *Challenger* and *Columbia*, seemingly tiny details—a cracked O-ring, a dislodged piece of foam—caused terrible disasters.

This is why, individually and organizationally, we have the patience to sweat the small stuff even when—actually, especially when—pursuing major goals. We've learned the hardest way possible just how much little things matter.

* * *

The night before my first spacewalk in 2001, I was calm yet very conscious of the fact that I was about to do something I'd been dreaming of most of my life. STS-100 was my second space mission but the first time I'd ever had so much responsibility for such a crucial task on orbit—I was EV1, the lead spacewalker. I felt ready. I'd spent years studying and training. Still, I wanted to feel even more ready, so I spent a few hours polishing the visor of my spacesuit so my breath wouldn't fog it up, unpacking and checking each piece of gear I'd need for the spacewalk, pre-assembling as much of it as I could and carefully attaching it to the Shuttle wall with Velcro—then double- and triple-checking my work, all the while mentally rehearsing the procedures I'd learned in the pool in Houston.

Scott Parazynski and I had been training for a year and a half to install Canadarm2, the robotic arm that would build the ISS, then in its infancy. In May 2001, the Station was just a fraction of its current size; the first parts of the ISS had only been sent into orbit three years earlier, and the first crew took up residence in 2000. Our crew hadn't even been inside the Station yet. We'd docked *Endeavour* to it a few days before but hadn't yet been able to open the hatch because our EVA was going to take place from the Shuttle airlock—a depressurized bridge, in essence, between the two spacecraft.

That night I felt a little like a kid on Christmas Eve. I wanted to get to sleep right away, to make the morning come faster. The visuals, however, were more appropriate to Halloween: on the Shuttle we slept in sleeping bags tethered to the walls and ceiling, an oddly macabre den of human chrysalises, hovering and still. I woke in the night and checked the green light of my Omega Speedmaster astronaut watch. Hours to go. Everyone else was fast asleep. I fell back asleep too until, with a burst of static, the small speaker in the Shuttle middeck erupted with wake-up music from Houston, a song Helene had chosen for me: "Northwest Passage" by Stan Rogers, one of my favorite folk singers. I slipped carefully out of my sleeping bag, found the microphone, said thanks to my family and everyone at Mission Control and started to get ready to go outside.

There are multiple, sequential, vital steps to follow for an EVA—mess one up, and you won't make it out of the spaceship. It would be many busy hours until Scott and I could float out of the airlock and NASA had choreographed them down to five-minute slices, even dictating when and what to eat for breakfast: PowerBars and rehydrated grapefruit juice. I shaved, washed up, used the toilet—I didn't want to have to use my diaper if I could possibly help it. Then I pulled on the liquid cooling garment, which is like long underwear with a lot of personality; it's full of clear plastic tubing that water flows through, and we can control the temperature. It feels stiff, like a cheap Halloween costume, but that doesn't matter when you're outside: when the sun is shining on you during a spacewalk, the fabric of the spacesuit gets extremely hot and personal air conditioning seems like a fine idea.

About four hours later, Scott and I were finally floating head to toe in our spacesuits, carefully and slowly depressurizing the airlock and checking and rechecking the LED displays on our suits to make sure they were functioning properly and could keep us alive in the vacuum of space. If we got out there and somehow there was a leak in the suit, our lungs would rupture, our eardrums would burst, our saliva, sweat and tears would boil and we'd get the bends. The only good news is that within 10 to 15 seconds, we'd lose consciousness. Lack of oxygen to the brain is what would finish us off.

Bobbing gently in the airlock, though, I am not pondering my demise. This is the restful part of the day, a little like the point in a cross-country flight when you look out the window and see Nebraska. We will be busy again at some future point, but now, we are in limbo, still hooked to the ship by our umbilicals, anaconda-like hoses providing cooling, oxygen, comms and power.

When the airlock has finally depressurized, I grab the handle on the hatch and turn it—not easily, because nothing in a spacesuit is all that easy. I talk calmly to Houston as I turn, but when it clicks into place and I feel the hatch move, I think, "Phew, it's opening." On a previous mission the handle had jammed, just locked up completely, and the astronauts had had to give up and go back into the Shuttle. The hatch itself is almost like a manhole, and it has to be removed and stowed in a bike-rack-like contraption overhead. I still can't see outside, because of the white fabric insulating cover over the opening, but suddenly the airlock is brighter, bathed in muffled sunshine. Once I stow the fabric cover I am looking at the payload bay of the Shuttle itself, with just a sliver of the universe in my field of view. Of course all I want to do is get out there, but detaching the umbilical is a production: you have to do it really carefully because the connectors are fragile, then shroud it and mount it securely to the wall so it's ready in case you have to race back into the airlock to stay alive.

Time to go out. Oh. The square astronaut, round hole dilemma. My exit will not be graceful. But my number one concern at this point is to avoid floating off into space, so just as we've been taught, I'm tethered to Scott, who is attached to the structure, and I'm holding another tether to attach to the rail mounted on the side of the Shuttle. I lower the gold shield on my visor to protect my eyes from the sun and carefully, gingerly, wriggle my bulky, square, suited self out of the airlock. I'm still inside the belly of the beast, in the payload bay, but my suit has become my own personal spaceship, responsible for keeping me alive. Emerging from the bay, my existence narrows to a single point of focus: attaching my tether to the braided wire strung from one end of the vehicle to another. I lock onto that and tell everyone I'm securely tethered. Now Scott can detach inside and come join me. Waiting for him I check behind me, to be sure I haven't accidentally activated my backup tank of oxygen, and that's when I notice the universe. The scale is graphically shocking. The colors, too. The incongruity is stupefying: there I was, inside a small box, but now—how is this possible?

What's coming out of my mouth is a single word: Wow. Only, elongated: Wwwoooowww. But my mind is racing, trying to understand and articulate what I'm seeing, to find analogies for an experience that is so unique. It's like this, I think. It's like being engrossed in cleaning a pane of glass, then you look over your shoulder and realize you're hanging off the side of the Empire State Building, Manhattan sprawled vividly beneath and around you. Intellectually, I'd known I was venturing out into space yet still the sight of it shocked me, profoundly. In a spacesuit, you're not aware of taste, smell, touch. The only sounds you hear are your own breathing and, through the headset, disembodied voices. You're in a self-contained bubble, cut off, then you look up from your task and the universe rudely slaps you in the face. It's overpowering, visually, and no other senses warn you that you're about to be attacked by raw beauty.

Another analogy: Imagine you're in your living room, intently reading a book, and then you look up casually and you're face to face with a tiger. No warning, no sound or smell, just suddenly, that feral presence. There was something similarly surreal and dreamlike about the sight in front of me now, which I couldn't reconcile with my prosaic fumbling with the tether hook a moment before. Of course I'd peered out the Shuttle windows at the world, but I understood now that I hadn't seen it, not really. Holding onto the side of a spaceship that's moving around the Earth at 17,500 miles an hour, I could truly *see* the astonishing beauty of our planet, the infinite textures and colors. On the other side of me, the black velvet bucket of space, brimming with stars. It's vast and overwhelming, this visual immersion, and I could drink it in forever—only here's Scott, out of the airlock, floating over toward me. We get to work.

After about five hours, the installation is going well, albeit a little slowly, when I suddenly become aware that droplets of water are floating around inside my helmet. An EVA is incredibly taxing, physically, and over the years we've tried putting some sort of food, a Fruit Roll-Up or something like that, inside the suit so that at least we have something to eat. But we've never figured out how to make food work, it's been messy and more of a hindrance than a help, so typically we just have a water bag. You bite on the straw to open a little valve at one end, then suck out water—hypothetically, anyway. My water bag hadn't worked right since we started the EVA and now it was apparently leaking. Great.

I'm trying to ignore these little globs of water floating around in front of my face when suddenly my left eye starts stinging. Wickedly. It feels like a large piece of grit has been smashed into my eye and instinctively, I reach up to rub it—and my hand smacks into the visor of my helmet. "You're in a spacesuit, moron!" I remind myself under my breath. I try blinking repeatedly and whipping my head hard from side to side to try to dislodge whatever it is, but my eye keeps stinging and won't stay open for more than a blurry second before snapping shut again.

We've trained for many eventualities during an EVA, but partial blindness was not one of them. So what to do? Well, take stock: I'm tightening the bolts on Canadarm2 using a big handheld drill. My feet are clicked into place in the foot restraints and my tether is firmly attached to the Station; I'm at no imminent risk. The rest of my senses are fine and I've still got one good eye. I decide to keep working and tell no one. So I move on to the next bolt and start torquing it into place. My left eye, however, is now not only smarting but actually filling with tears.

Tears need gravity. On Earth, a little duct above your eye generates tears that flush out whatever irritant is in your eye and then overflow down your cheek and drain down your tear duct, making your nose run. But in weightlessness, tears don't flow downward. They just sit there and, as you keep on crying, a bigger and bigger ball of salty liquid accumulates to form a wobbly bubble on your eyeball.

Now for some key anatomy. My great-grandparents were all from northern England and southern Scotland, and while Yorkshiremen and Scots are noted for their toughness and stoicism, they are not remarkable for their noses. Instead of a proud, protruding hawk-like nose, they bequeathed me a more humble bridge, which the growing ball of tears in my left eye easily oozes over, like a burst dam, promptly invading my right.

Which also snaps shut, because whatever's contaminating the left eye hasn't been diluted by my tears so now my right eye is tearing heavily, too. I try to force my eyes open, but there's not much point—all I can see is a watery blur before my reflexes kick in and my eyelids close. In the space of just a few minutes, I've gone from 20/20 vision to blind. In space. Holding a drill.

"Houston, EV1. I have a problem." As the words come out of my mouth I can easily picture the reaction on the ground, having CAPCOMed so many flights myself. First there will be concern for me personally, and then, seconds later, everyone at Mission Control will be galvanized: people will start tossing out theories about causes, wondering aloud what this means operationally and trying to figure out solutions.

To Scott and me, underreacting seems like the best option: I can't see, but he's just fine and still working away on the wiring on another part of the Station. Pointless for him to stop and make his way over, as there's absolutely nothing he can do for me. Of course, if it turns out there's no way to solve my problem, he'll have to lead me back to the airlock and get me inside safely, but we both agree we're not at that point yet. Nor do I want to get there. I need to get this job done, and my country is counting on me; the Canadian-designed and -built Canadarm2 is both a test and dazzling proof of our robotics capability. The EVA itself is also a big deal back home, because no Canadian has ever walked in space before. In other words, it's really not a good time to be having eye trouble.

Fortunately, the flight director is Phil Engelauf, who knows me well. I've worked alongside him many times as CAPCOM for Shuttle flights, and he's willing to cut me a lot of slack instead of ordering me back inside, pronto. He lets me sit tight for a bit while people scramble to figure out just how much danger I'm in. I know the ground is abuzz because every time the CAPCOM speaks to me, I can hear the hubbub in the background: *How did this happen? Is it going to get worse? What can we do?* It's not insignificant that the arm is only partially attached—yes, crew safety is the number one priority, but we can't just leave this vital piece of equipment flapping off the side of the Station.

After a few minutes, the focus on the ground is figuring out what's causing the contamination. This being the space business, they go straight to the worst-case scenario: maybe the problem is related to the air purification system in the spacewalking suit, which relies on lithium hydroxide to remove carbon dioxide. Lithium hydroxide is really caustic and can severely damage your lungs; eye irritation is one of the first signs that there's been a leak. So maybe I'm experiencing early symptoms of lithium hydroxide exposure and I've only got a couple more minutes to live. The CAPCOM, Ellen Ochoa (now the director of the Johnson Space Center), calmly tells me to open my purge valve—essentially, to open a hole in my suit and start flushing out the potentially contaminated air I've been breathing until it's all gone or at least highly diluted by the fresh oxygen being pumped into my suit.

This goes against my survival instincts, but, okay. I open my purge valve—luckily, I've practiced so many times that I can reach up beside my left ear and open it with no problem and no eyesight—and start dumping my air into space. So now I'm blind, listening to a hissing noise as my oxygen merrily burbles out into the universe. It's a curiously peaceful moment. Spacewalking is largely a visual experience; your other senses are barely stimulated. The brilliant colors of the Earth, the shining reflections off the spaceship and the profound blackness of space are what confirm to you where you are. Without sight, my body is telling me that nothing at all out of the ordinary is going on. I feel more like I'm under the covers at home in bed, dreaming about the Space Station, than hanging onto the side of it, in mortal danger.

My CAPCOM is listening to the medical doctors, the biomedical engineers, everyone who's working away at Mission Control, but she says, as though we're just having a pleasant conversation, "So Chris, we're looking at all the data, where your oxygen pressure is at right now. How are you feeling?" Weirdly enough, I'm feeling unconcerned, because Scott is out here with me. He's a physician and a commercial pilot and a mountain climber, and I've never met anyone who can outwork him: the guy's mind and body just never stop. Plus, I'm still breathing, a lot of good people are working the problem and I'm certain I'm not going to die in the next 60 seconds. The fact that I'm not coughing makes me reasonably confident there hasn't been a lithium hydroxide leak. I have to let the people on the ground do their job, and purge my oxygen as a precaution, but I've already decided I'm not going to let this go on too long. The suit has a significant amount of oxygen, enough for eight or even ten hours, and I also have a secondary O2 tank, so I can bleed out oxygen and stay alive for a long, long time. But I need to get back to work, and who knows how much longer we'll have to be outside to finish attaching the robot arm.

Actually, I'm getting antsy: we're wasting time here. I'm contributing absolutely nothing to the project I've come to space to do. So I start trying everything I can think of to un-blind myself: shaking my head around to try to brush my eyes against something in the helmet, blinking for all I'm worth. I know the doctors are undoubtedly telling Phil, "We've got to bring him inside right this minute and figure out what's going on." So I say, "Know what? I feel no lung irritation at all and I think my eyes are starting to clear a little bit." It's even sort of true. My eyes are still killing but I feel marginally less sightless.

I ask if I can stop purging oxygen and Phil agrees. Meanwhile, I keep blinking and blinking and blinking, and thankfully, 20 minutes on, I can now see a little bit. Sure my eyes still sting and everything looks a bit cloudy, but a couple more minutes pass and I think I can see well enough to continue installing the arm. I tell the ground I'm ready to get back to work. Happily, the response is, "All right, you're the one there and you know best what the situation is." In the meantime, Mission Control instructs the crew inside the Shuttle to get the medical gear ready so that when I come back in they can sample my tears and the crusting around my eyes to try to figure out what the problem is.

In the end, Mission Control wound up letting us go long on our spacewalk, which was scheduled to last six and a half hours. The vast majority of spacewalks are seven hours or less, but because Scott and I were both telling the ground how well we were doing, we were allowed to stay out almost eight, to try to get everything done.

Nearing the end, I look down to watch the world pour by. Having overcome this obstacle and knowing the two of us have got everything buttoned up right and have accomplished what we set out to accomplish—it's a big moment. But with a spacewalk, the very last step is as important as the first one, so not until we've repressurized the airlock and are actually back inside our spaceship do I let myself relax. As soon as I do, I feel completely drained and just float limply, shivering with cold. My body is out of fuel. But when one of the crew medical officers floats over with a 3-foot-long cotton swab he's fashioned out of stuff he found on board and tells me he'll be jabbing this thing in my eye to take samples, I do still have enough energy to laugh.

Later, discussing what went wrong, we all suspected the droplets from my water bag—maybe they'd mixed with a bead of sweat, or something from my hair, or something inside the suit itself. We were going over all the possibilities with Mission Control when the CAPCOM asked, "Chris, did you remember to use your anti-fog stuff?" Of course I had. The night before I'd polished the visor of the suit so it wouldn't fog up like a ski mask. "Well, we think you didn't do it perfectly. Probably you didn't get it all off." Apparently the solution is basically dishwashing detergent; mix it with a few droplets of loose water and it's as though you've squirted soap directly into your eye. My first response to this news was, "Really? We're using *detergent*? No More Tears baby shampoo wasn't an option?"

But my second response was, "Next time, I'll be even more detail-oriented." A spacewalk with a multi-million-dollar piece of equipment that was—is—absolutely vital to the construction of the ISS was jeopardized because of a microscopic drop of cleaning solution.

When I went out for my next EVA two days later, I wiped off my visor so vigorously that I'm surprised I didn't rub right through the thing. Eventually, NASA changed the solution we use to clean our visors to something a little less noxious. But in the meantime, thanks to my widely publicized oversight, all astronauts knew to be fanatical about wiping down the interior of their visors. And when a couple of them also wound up temporarily blinded on their own spacewalks, Mission Control knew what the trouble was: "Remember Hadfield? It's the anti-fog solution."

That's why it's so worth it to sweat the small stuff. And even in my line of work, it's all small stuff.

5
THE LAST PEOPLE IN THE WORLD

THERE'S NO SUCH THING as an accidental astronaut. On average, new astronauts are 34 years old; wanting the job has driven their choices for many years. The odds of being selected are now slimmer than ever. During the last recruitment in Canada in 2009, just two astronauts were chosen from a field of 5,351 applicants; that year NASA reviewed 3,564 applications for nine spots. The selection process is both rigorous and invasive. A Ph.D. is just table stakes, and a nose polyp is a deal breaker; applicants who make the final rounds are subjected to psych tests, rectal exams, and endless interviews and written tests. People who are willing to put themselves through all that are, by definition, highly competitive.

I know I was when I applied for the job. I didn't think being selected was a sure thing—far from it; the process was nerve-wracking—but I was confident that I was a good fighter and test pilot. Being chosen as one of four new CSA astronauts felt like the biggest possible affirmation of my competence, and I was both proud and excited when shortly thereafter, I was told to pack my bags and head to Houston along with Marc Garneau to start training as a member of the class of 1992. It was the heyday of the

Shuttle era, so ours was a big class by current standards: 24 of us in all. We took the elevator up to the Astronaut Office at JSC, quietly giddy: this was one of the hardest offices to join in the world, yet we'd made it. We were the *crème de la crème*.

Then we got off the elevator.

Just like that, we were nobodies. We weren't even called astronauts but ASCANs (pronounced exactly as you might imagine), meaning "astronaut candidates." Plebes. No hazing was required to knock us down a peg. Just looking around the office and seeing people we'd idolized for years did the trick. When I was assigned to a desk beside John Young—one of the original Gemini astronauts, one of only a dozen men to walk on the Moon and the commander of the very first Space Shuttle flight—I didn't feel like I'd finally arrived. I felt like a gnat.

In the course of my first day at JSC I went from the peak of my profession to the bottom of the food chain—and I was down there with a bunch of other overachievers who were used to being on top and determined to get back there ASAP. It's not as though there wasn't camaraderie. There was. Each class has its own particular character and nickname: members of a particularly large class were "the sardines," and those of us who joined in 1992 were called "the hogs" (partly thanks to a Muppets skit called "Pigs in Space," and partly because we decided early on to sponsor a pot-bellied pig at the Houston Zoo). There was definitely a sense that we were all in this together, but the environment was also highly competitive, without the competition ever being explicitly acknowledged. Each of us was being evaluated and compared on everything we did—*everything*—and it was very clear that space flight assignments would be based on how well we performed. So the demands were bottomless. I never wanted to turn down any request or opportunity, and like everyone else, I kept trying to make it all look easy.

In the meantime, my family had relocated to Houston, which meant a new house, new schools for the kids and for Helene, a new job. The first year is very tough on families because of all the adjustments and changes. Some ASCANs' marriages implode, partly because of the strain on the spouses but largely, I think, because of the astronauts' struggle to adapt to a new place in the pecking order. The reasoning seems to go like this: *My dream's come true, yet I feel like a gnat—but I'm still the same high achiever, so the problem must be . . . my marriage!* I'm extremely lucky because my family approached our many moves with a sense of adventure. Still, coming from the military, we found the whole set-up in Houston a little disconcerting at first. It seemed military, yet it wasn't. Typically on a squadron, pilots' families live near each other on base and tend to do things together, too. But at NASA, everybody's just too busy. Having grown accustomed to a certain communal rhythm, it felt lonely for all of us at first.

In a sense, too, going to work every day was disconcerting. During the year I was an ASCAN, the learning curve was daunting and there weren't a whole lot of opportunities to stand out. After that first year, I worked on certifying payloads, which involved endless meetings to make sure that all the science experiments were actually safe for space flight. In the meantime, just like all my classmates, I was going through general training: geology, meteorology, orbital mechanics, robotics and so on. People who'd been in the Astronaut Office only a year or two longer seemed to be light years ahead, even though they hadn't been to space yet.

Then came the day when the first person in our class got assigned to a space flight. It was a great moment: "Wow, one of us made it!" It felt like a group affirmation, as though all of us were on our way at last. Then the second person got assigned, and it wasn't me. I told myself, "Okay, they picked a scientist—they weren't looking for a pilot." Then in the middle of that night: "I'm Canadian. That's probably why they didn't pick me." Then the third person got assigned, and the fourth, and I started thinking, "What's wrong with me? I've always been good at stuff. Why am I not getting assigned?"

This is when attitude really started to matter. I have a clear memory of giving myself a pep talk right about then that started with, "Don't be an idiot." I reminded myself that I wasn't sitting around doing nothing. I was learning so much every day that I could almost hear my neurons firing.

If you've always felt like you've been successful, though, it's hard not to fret when you're being surpassed. The astronauts who seem to have the hardest time with it are, interestingly enough, often the ones who are most naturally talented. Just as some people can pick up a golf club for the first time and play incredibly well, some astronauts are simply more gifted than the rest of us. They have great hands and feet—the first time they got in a plane, they could fly as well as or better than the instructor. Or they're academic superstars with dazzling interpersonal skills. Whatever their particular combination of gifts, they're standouts, and until they got to JSC, everything was easy: they won the flying competitions, aced all the tests, told the best stories—all without breaking a sweat.

Early success is a terrible teacher. You're essentially being rewarded for a lack of preparation, so when you find yourself in a situation where you *must* prepare, you can't do it. You don't know how.

Even the most gifted person in the world will, at some point during astronaut training, cross a threshold where it's no longer possible to wing it. The volume of complex information and skills to be mastered is simply too great to be able to figure it all out on the fly. Some get to this break point and realize they can't continue to rely on raw talent—they need to buckle down and study. Others never quite seem to figure that out and, in true tortoise-and-hare fashion, find themselves in a place they never expected to be: the back of the pack. They don't know how to push themselves to the point of discomfort and beyond. Typically, they also don't recognize their own weaknesses and are therefore reluctant to accept responsibility when things don't turn out well. They're not people you want on your crew when you're laboring in wicked environmental conditions with very specialized equipment and a long list of goals to accomplish in a short period of time. They go from being considered rock stars to having a reputation as people you can't count on when things are going badly.

* * *

There's a big variety in terms of ability and skill within the astronaut corps, more than most people imagine, though much less than there used to be back when 50 people were flying a year and the crews were larger, so everyone didn't have to be good at everything. On the Shuttle you really only needed two people who were good robotic operators. Today, with a crew of just three on the Soyuz, at least one of whom is a cosmonaut, if you're not good at robotics and not qualified for EVA, you're likely not going to be assigned.

When the missions were just two weeks long, crews were put together a little bit like a sports team: it was all about the mix. Administrators wanted both experienced people and rookies and sought a balance between military and academic types, in-your-face people and laid-back, affable ones. Of course, politics played into it too: whose turn it was to fly mattered sometimes, as did nationality. Canadians were not usually high up on the list, but when Canadarm2 was being installed, it made sense for one of us to go. Some crews never really did jell, but it wasn't all that important. If you're only off Earth for a couple of weeks, you can put up with just about anyone. You don't need to have the time of your life. You need to get the job done.

On the ISS, by contrast, homogeneity has a greater value because you need redundancy of skills—if only one of the three astronauts on board has medical training and she's incapacitated and in dire need of medical care herself, you've got a serious problem. Training is also much more solitary. For two years astronauts are mostly solo, training and studying one-on-one with instructors, and then, in the last six months before a flight, when everyone has the requisite skills, we start to integrate as a crew.

Sometimes integration is not so easy, because we don't get to pick our fellow travelers. It's like a shotgun wedding, minus the conjugal rights—and the "honeymoon" is half a year in isolation, where we have to be able to count on one another for absolutely everything: companionship, survival, taking responsibility for a fair share of the work.

That's why "Who are you flying with?" is the first question astronauts ask each other. No one wants to go to space with a jerk. But at some point, you just have to accept the people in your crew, stop wishing you were flying with Neil Armstrong, and start figuring out how your crewmates' strengths and weaknesses mesh with your own. You can't change the bricks, and together, you still have to build a wall.

Sometimes you get lucky. Both Tom Marshburn and Roman Romanenko, my crewmates on my last mission, have superlative technical skills as well as a killer work ethic. They are also two of the most easygoing and pleasant people on or off the planet. I didn't have to make peace with the fact that I was going to space with them. I had to refrain from crowing about my good fortune.

The longer the flight, the more important personalities become. If the three of you don't get along on Earth, you're even less likely to be able to tolerate each other after a few months without the benefit of showers. Or Scotch. Some of the first American astronauts who went to Mir for long stays experienced depression and felt isolated and irritated both by crewmates and by what they felt was a lack of support from Mission Control. When you can't even go outside to let off steam, personality conflicts can compromise a mission or derail it altogether. Simmering tensions have boiled over in the past, according to some of the first long-duration cosmonauts, who have colorful stories of personality clashes. I've heard rumors of fistfights and refusing to speak to one another (and the ground) for days on end. So these days, NASA looks for a certain type of person, someone who plays well with others.

One thing hasn't changed, though: astronauts are, without exception, extremely competitive. I may have mentioned this before. So how do you take a group of hyper-competitive people and get them to hyper-cooperate, to the point where they seek opportunities to help one another shine?

It's a bit like gathering a group of sprinters and telling them that, effective immediately, they'll be running an eternal relay. They've still got to run as fast as they can, only now, they've got to root for their teammates to run even faster. They have to figure out how to hand off the baton smoothly so that the next person in line has an even better shot at success than they did.

For some astronauts, the transition is relatively painless—a relief, even, after decades of solitary striving. For others, it's a huge shock to the system and requires a fundamental reorientation.

I was somewhere in the middle. To my chagrin, I was the kind of father who rarely let my kids win—they had to earn victory, fair and square. I don't have a lot of regrets in life, but one of my biggest is that when my son Kyle was about 10 and was proudly demonstrating how many laps he could swim underwater without taking a breath, I jumped in the pool and swam one more length than he did. It was an unthinking moment, and a great demonstration of the destructive power of competitiveness. I didn't just show up my child; I risked damaging his self-confidence and our bond.

Paradoxically, it took a few years working with other wildly competitive people for me to learn to think of success as a team sport. To instill and reinforce expeditionary behavior—essentially, the ability to work in a team productively and cheerfully in tough conditions—astronauts do survival training, on water and on land. Over the years I've done that with the U.S. and the Canadian military, as well as participating in wilderness expeditions in Utah and Wyoming, both run by the National Outdoor Leadership School (NOLS). The specifics of the experiences were different, but the focus was always the same: figuring out how to thrive, not just individually but as a group, when you're far outside your comfort zone.

Survival training simulates some aspects of space travel really well. In both cases, a small group of people is thrown into a challenging environment with specific objectives to accomplish and no one else to rely on except each other. At NOLS, for instance, we divided into teams and took turns as leader, with the goal of safely traversing a wilderness route in 10 to 14 days. It was a harsh collective experience: sleeping rough, orienteering, rappelling down cliffs, searching for pure water and so forth, all while lugging a heavy backpack.

During the Utah course, I remember reaching the top of an especially daunting ridge and looking down to the valley where we were supposed to set up camp for the night. Our hearts sank. There was no way to get down there. Everyone was tired and cranky, and had there been an option to quit the course and be airlifted to the nearest Hilton, I think most of us would have signed up for it on the spot. But after studying the situation, Scott "Doc" Horowitz and I thought it just might be possible to descend by zigzagging down a particular slope. If we were wrong, though, the group might get stuck there as night fell and temperatures plummeted; we'd be in far more danger on a steep, rocky slope than we currently were at the top of the ridge. So instead of trying to persuade everyone else to try our route, Scott and I volunteered to scout it out. We proved to ourselves that it was doable, then climbed back up to show the others how to descend safely. The lesson: good leadership means leading the way, not hectoring other people to do things your way. Bullying, bickering and competing for dominance are, even in a low-risk situation, excellent ways to destroy morale and diminish productivity. A few NASA teams have in fact come somewhat unglued and been unable to complete survival exercises, which no doubt was noted back at JSC by the people who determine flight assignments.

Another thing we learned in survival training is that risk management is crucial when you're in the middle of nowhere. I was extremely careful scouting our descent because I knew that if I broke my ankle, I wouldn't be seen as a hero or a martyr. I'd be the guy who compromised the mission. Groupthink is a good thing when it comes to risks. If you're only thinking about yourself, you can't see the whole picture. Whether in the mountains of Utah or clinging to the outside of the ISS, getting hurt—or losing the only hammer the group has, or rushing through a tricky procedure—creates serious problems for the entire team.

For me, the takeaway from all my survival training is that the key question to ask when you're part of a team, whether on Earth or in space, is, "How can I help us get where we need to go?" You don't need to be a superhero. Empathy and a sense of humor are often more important, as I was reminded during the most arduous survival training I ever did, in central Quebec with five other astronauts. We were on the edge of the Laurentians, so the terrain was mountainous and hiking would have been challenging at the best of times, but this was February and the snow was relentless. It just never stopped coming, almost a foot a day, and for two weeks we had to trudge through the drifts in snowshoes to break a trail for the sled that was loaded down with our food and supplies. When you think of a sled, you probably picture flying down a hill. That was not our experience. This one weighed 300 pounds and wasn't going anywhere unless we were pushing and pulling it. We took turns at the front, a few of us at a time, straining to drag this thing, often uphill. We'd go 15 paces, then, so exhausted we were almost spitting blood, take a break and trade places with the people who'd been pushing. I was the only Canadian, so I was supposed to be used to rugged winter outings of this nature, but . . . I wasn't. I didn't grow up in the wilds, sleeping in snowdrifts.

The situation was perfect for developing leadership—and followership—skills, and it was a great test of physical endurance and mental stamina, too. In retrospect, in fact, there's a pleasing, epic quality to the whole enterprise: the blinding snow, the heavily laden sled, the laborious slogging. At the time, though, it didn't feel pleasing at all.

This is where expeditionary behavior comes in. You can choose to wallow in misery, or you can focus on what's best for the group (hint: it's never misery). In my experience, searching for ways to lighten the mood is never a waste of time, particularly not when it's 10 degrees below zero. Among our supplies there was a pineapple, oddly enough, and someone came up with the idea of carving a face on it and calling it Wilson, in homage to the volleyball that is Tom Hanks's only companion when he is stranded on a tropical island after a plane crash in *Cast Away*. Wilson became a member of our crew and was treated with the same reverence Hanks showed his volleyball, right up until the pineapple turned a rather unsavory color and a funeral was deemed necessary. But Wilson served his purpose, morale-wise.

I hit on something during that Quebec expedition that I've used subsequently as a distraction when the going gets tough: suggesting that one by one, we each describe how we got engaged to our spouses. Everyone liked telling his or her own version. I liked hearing other people's stories, too, because most of the other astronauts were older than I was when they got engaged, and their proposals were considerably better orchestrated than my own. I asked Helene to marry me on Valentine's Day. I was 21, still in military college, and took her out for a candlelit dinner with the ring in my pocket, planning to propose in the restaurant. But once we got there, it just didn't feel right, so I wound up asking her later that evening, sitting on the side of a bed in the Holiday Inn in Kingston, Ontario. I was nervous, she cried and neither of us remembers exactly what was said, though Helene's recollection is that the proposal would have benefited from a poetic flourish or two. Sharing that story with the other astronauts on the survival course gave them insight into my life, and their own tales of picture-perfect proposals on sunlit beaches, complete with beautifully crafted speeches, gave me insight into theirs. Storytelling also provided a pleasant and prolonged diversion from the Sisyphean task of dragging that sled through the snow.

That was the second hardest experience of my life, physically. The first was when I was about 14 and, along with the rest of my family, had spent a long late-summer day in the fields, harvesting corn. We were just sitting down to dinner when my dad came in after sticking a long thermometer into one of the storage bins to check that the dried kernels of corn weren't heating up and starting to ferment. Well, they were, and if we didn't do something fast, we were going to lose the farm's entire profit for the year. So we all got up from the table and ran out to the barn and started shoveling the corn, continuously moving it from the bottom of a 6-foot-deep bin to the top, to aerate and cool it. My whole family worked through the night to save the crop. There was no question of stopping.

Or of whining. My dad could be a stern taskmaster and on principle didn't believe children should complain, but he also disapproved of whining because he understood that it is contagious and destructive. Comparing notes on how unfair or difficult or ridiculous something is does promote bonding—and sometimes that's why griping continues, because it's reinforcing an us-against-the-world feeling. Very quickly, though, the warmth of unity morphs to the sourness of resentment, which makes hardships seem even more intolerable and doesn't help get the job done. Whining is the antithesis of expeditionary behavior, which is all about rallying the troops around a common goal.

It's easy to do that in an event-driven situation, like a Shuttle mission that's built around repairing a telescope or installing new equipment on the ISS. When the objective is well defined and time-limited, most people can stay focused on achieving it. On the ISS, however, the goals are fuzzier: keep the experiments going, maintain the Station. There are a lot of finicky, janitorial-type tasks, and as with housework, you never really finish. Plus, we're there long enough for petty grievances and irritations to accumulate and to seem to mount in importance, too. So as commander of Expedition 35, I deliberately discouraged whining whenever I noticed it creeping into conversation. However, I couldn't simply impose my will on the rest of the crew. Only the crew's own appreciation of the value of expeditionary behavior made it possible for us to become a complaint-free group.

Each of them also made a point of promoting team spirit. Tom, for instance, is a medical doctor by training, and he has the ultimate gentle, supportive bedside manner. If he sensed that anyone needed help, he'd stop whatever he was doing and assist in a way that suggested that helping us out was really what he would rather be doing. He made us feel we were doing him a service, somehow, by allowing him to bail us out. Roman is one of those cheerful people who always seems on the verge of bursting out laughing. He understands the necessity of having fun and, if spirits flagged, lightened the mood by grabbing his harmonica or the Station's guitar and playing a riff from something we all knew.

On board the ISS there's a sack of holiday things: a small Christmas tree and lights, plastic Easter eggs, New Year's noisemakers, an assortment of party hats and so forth. This stuff has accumulated gradually over the years and provides an interesting, informal archeological record of crews past, but I mention it because Roman was always digging into that sack. On the way to video conferences with family or friends, or to record a greeting for someone, or to one of our group dinners, he'd put on a crazy orange jacket and Groucho Marx glasses—anything that would make him look ridiculous and get people laughing. He was also forever taking newly acquired English vernacular and applying it in ways he knew were ridiculous. Once we were working with a tricky piece of equipment that needed to be jiggled a little bit, and he suddenly instructed, in a strong Russian accent, "Shake what your mama gave ya!" then dissolved into laughter.

* * *

I've worked with some difficult people, too. One particularly abrasive astronaut flew on several Shuttle flights for which I was lead CAPCOM; we had to work together closely, particularly during the mission he commanded. The CAPCOM is the crew's trusted representative on the ground, and I really enjoyed trying to make sure things went smoothly for the crew—except when I had to work with this guy. He was highly skilled, technically, but also arrogant and confrontational, the kind of person who regularly swore at me, berated me and told me in no uncertain terms that I was a bumbling fool. I started to dread interacting with him, and when he dressed me down in front of Mission Control, I wanted to lash back, make my case in a legal manner, enlist supporters and try to convince them I'd done nothing wrong—everything about him just rubbed me the wrong way, professionally and personally.

Then I realized: Wow, he's really effective. This is his way of competing—trying to terrify and belittle others. His objective is to have a negative impact, and it's working. He's actually making me doubt my own competence.

Figuring that out helped me stop reacting emotionally to his abuse and start trying to figure out how to make the best of the situation. I quickly realized that I shouldn't take the guy's behavior personally. I was just one of hundreds of support people he thought were plotting his downfall; he reduced the secretary to tears on an almost daily basis. But even though I didn't have a lot of respect for him as a person, I was his junior and had to respect his role, whether he respected mine or not. I decided I had to let his criticisms slide by. So I did. I even reached a point of detachment where I was able to see clearly that he was a top operator of a complex vehicle who had some great skills and some fundamental problems. The trick to working well with him was to understand that the problems were his, not mine, and they all seemed to stem from his insecurity. He was unable to view his colleagues as anything other than competitors out to destroy him, who therefore needed to be squashed like bugs.

Once, flying up to Washington in a NASA jet, I stopped to refuel and a military guy I'd never met before noticed the plane and said, "Hey, do you know ____? What an asshole!" It was striking: of all the things he might have said to me on first meeting, his low opinion of that astronaut was the most pressing. I just said, "Wow. You've met him."

That incident really stuck with me. I would be horrified if a stranger met one of my colleagues and said, "Hey, do you know Chris Hadfield? I ran into him once. What a jerk!" I would be even more horrified if one of my colleagues, someone who knew me really well, heartily agreed.

It was a happy day for me when that astronaut left the office, but in retrospect, I learned a lot from him. For example, that if you need to make a strong criticism, it's a bad idea to lash out wildly; be surgical, pinpoint the problem rather than attack the person. Never ridicule a colleague, even with an offhand remark, no matter how tempting it is or how hilarious the laugh line. The more senior you are, the greater the impact your flippant comment will have. Don't snap at the people who work with you. When you see red, count to 10.

These are good rules in general but particularly in the space business. If I got into serious trouble on orbit—a medical emergency, say, or a catastrophic equipment failure—my crewmates would be my only hope of survival. For all intents and purposes, they'd be the last people in the world. That's a thought I try to keep in the forefront of my mind every day, not just in space but on Earth.

* * *

If your crewmates are the last people in the world, they're also the last ones you want to alienate or irritate. I grew up in a farmhouse with four brothers and sisters, so I've had a lot of firsthand lessons about the importance of consideration in tight quarters. But I needed another one, apparently, and got it during my last mission.

I'd been on the ISS about three weeks when I noticed my fingernails needed trimming. I'd never been in space that long before so hadn't faced this particular issue, and I knew that without gravity, dealing with the clippings might be tricky. So I came up with a really great idea: I'd cut my nails over an air duct intake filter. My new-guy idea was that every small clipping would get sucked right into the intake. It worked! I even recorded this improvisation on video so people on Earth could watch a mundane task made oddly interesting by the absence of gravity. I didn't think through all the implications, though. That weekend Kevin Ford, the commander of Expedition 34 and the person who was responsible for cleaning that part of the Station, undid the screws so he could vacuum behind the filter panel, thereby launching a hive of my dead fingernails into his face and everywhere else. He did his best to catch them all with the vacuum, but it couldn't have been pleasant. He came to me later and politely mentioned that next time I clipped my nails he'd appreciate it if I'd vacuum them off the intake immediately. I was mortified, but all I could do was apologize and make a note that the next time I felt smug about my cleverness, I should watch for the unintended consequences.

In the grand scheme of things, it was a minor mistake. But if I'd kept making more mistakes like that, it would have become a major irritant for everyone on board and ultimately, that could have chipped away at our effectiveness as a team. If you're seen as being consistently inconsiderate, or just out for yourself, there's a direct impact on communication and, usually, overall productivity. People simply won't work as well with you as they would with someone whose behavior was a little more expeditionary.

* * *

Over the years I've learned that investing in other people's success doesn't just make them more likely to enjoy working with me. It also improves my own chances of survival and success. The more each astronaut knows how to do, and the better he or she can do it, the better off I am, too.

For Expedition 34/35, my last mission, Roman was commander of the Soyuz, I was the left-seater, or co-pilot, and Tom was the right-seater. The Soyuz is designed to be flown by two people; the right-seater has no designated responsibilities beyond looking after himself or herself, so doesn't get detailed training. You could fly a suitcase in that seat, no problem. But Tom was eager to learn about the Soyuz, and to me, that seemed like a win-win proposition, both personally (he might wind up saving the day, noticing something Roman and I had missed) and in terms of our organization: the greater his depth of experience, the more valuable he would be to NASA post-flight. It took a little more of my own time and energy, training together after-hours and explaining procedures in detail, but it was a great investment, not just in him as an astronaut but in terms of his capability as a crewmate. Even in the sims, Tom could have the malfunction book open to the right page and point to the step Roman needed when something went wrong, or he could calculate how long the backup burn should be. If I'd said, "Look, Tom, just take care of yourself and we'll get you to the Station and back, no problem," our team would not have been as strong.

Having "overqualified" crewmates is a safety net for everyone, and I was lucky that Tom and Roman felt the same way and were willing to invest in my success, too. During training, when I messed up a Soyuz docking practical exam, Roman long-facedly commiserated with me, regaled me with stories about the times he and other cosmonauts had failed tests, suggested techniques and tactics I could use to improve my performance and then rejoiced with me when I retook the exam and passed. He did that not just because he's a nice person but because the higher my skill level, the more peace of mind *he* had. He wanted a crewmate who could, in an emergency, be of some use.

It's not enough to shelve your own competitive streak. You have to try, consciously, to help others succeed. Some people feel this is like shooting themselves in the foot—why aid someone else in creating a competitive advantage? I don't look at it that way. Helping someone else look good doesn't make me look worse. In fact, it often improves my own performance, particularly in stressful situations.

Once, I was doing water survival in the Black Sea where, in teams of three, we were simulating water landings in the Soyuz. The scenario was that we'd splashed down in the ocean and needed to get out of the capsule and into a life raft within half an hour, using the right techniques. I was doing this exercise with André Kuipers, an experienced astronaut who's as big as you can be and still fly on the Soyuz, and Max Ponamaryov, a small, strong cosmonaut in his late 20s who'd just completed introductory training. It was summertime, we were wearing pressure suits and it was hot in the capsule—so hot that each of us had had to swallow a transmitter so our core temperature could be monitored for safety. All of us were sweating like crazy and basically just wanted out of that tiny capsule ASAP. But first we had to take off our pressure suits—awkward even if you have all the room in the world—and put on water survival suits, which are a little like down-filled snowmobile suits, then pull waterproof gear over them. In other words, we had to get a whole lot more uncomfortable before we could get out of there.

Focusing on the discomfort, though, was only going to make it worse. Instead, we decided to focus on how to support each other and make Max's first experience as commander a big success. André, who's a medical doctor, kept reminding us to drink more water so we wouldn't get dehydrated, but Max, who doubtless felt some pressure as a rookie to prove how tough he was, was initially reluctant. So André and I started chugging water, which made it okay for Max to drink, too. Likewise, Max insisted on trading places with André, who despite being the biggest of us had been assigned the most cramped seat, the left one, and was having the most trouble getting out of his pressure suit. Just when the heat felt least bearable, I fake-shivered and said, "Brrr, it's cold!" It provided not only comic relief but, for whatever reason, a bit of physical relief as well, so we all started doing it and for a glorious moment or two almost believed we weren't bathed in sweat. André's water survival suit didn't fit but we helped him wriggle into it as best as he could, then did the egress properly and the end result was that Max emerged as a star commander.

Possibly we could have completed the sim in about the same amount of time if our attitude had been "every man for himself," or if André and I had taken charge because we had more experience. But I doubt it. I think focusing on helping Max deliver a win helped us tough out the physical unpleasantness and improved our performance individually, too. The other group doing the exercise couldn't get through the clothing swap and had to be rescued and get extra training the next day. The exercise really had a lot less to do with water survival than with deliberate teamwork.

It's counterintuitive, but I think it's true: promoting your colleagues' interests helps you stay competitive, even in a field where everyone is top-notch. And it's easy to do once you understand that you have a vested interest in your co-workers' success. In a crisis, you want them to want to help you survive and succeed, too. They may be the only people in the world who can.

6 WHAT'S THE NEXT THING THAT COULD KILL ME?

JUST AS IT'S MORE DANGEROUS to walk through a rough neighborhood alone, a military pilot is more vulnerable flying solo over enemy territory. That's why we learn to fly in formation: if you've got someone on your wing, you can keep an eye on one another.

However, you can also kill one another without too much effort. Flying in close formation requires laser-like focus; you have to be able to ignore absolutely everything aside from following the leader and executing the maneuvers precisely. The importance of this was driven home to me one of the very first times I ever flew in formation, during basic jet training. We were in our Tutors, four across, staggered back like fingertips; I was third in the row, boxed in, when I noticed something moving in my field of vision. Inside my visor, actually. A bug of some sort, so close to my eye that I had trouble making out exactly what it was.

Oh. A bee. A big one, a couple of inches from my eyeball.

It's not unheard of for an insect to get trapped in the cockpit when you close the canopy, but I'd never had one inside my visor before. And this bee was crawling slowly and woozily—groggy, probably, because of the thinness of the air at that altitude.

Disorientation might make it more defensive and more likely to sting, but there was nothing I could do about it. I couldn't blow on it because I was wearing a mask—nor did I want to do anything to startle it. The key thing was to continue to fly my plane steadily. In the middle of the line, I was stuck. I couldn't safely peel away with no warning. If I broke formation I'd endanger the pilots on either side of me. Our planes were that close together.

Knowing how high the stakes were helped me override my instinctive desire to put a lot more distance between me and that bee. I can't say that I was able to forget it was there—I had no choice but to look directly at the thing; shutting my eyes was not an option. But I did manage to hold formation until I had a chance to radio and ask the leader to let me fall back long enough to open my visor and lose the bee.

Nothing focuses your mind quite like flying a jet. That's one reason NASA requires that astronauts fly T-38s: it forces us to concentrate and prioritize in some of the same ways we need to in a rocket ship. Although simulators are great for building step-by-step knowledge of a procedure, the worst thing that can happen in a sim is that you get a bad grade on your performance. In a T-38, an old training plane that's fast but short on fuel and not all that responsive, you have to operate complex, unforgiving systems in a dynamic environment; the weather and winds are always changing. You're constantly forced to make judgment calls, like whether to turn back or push on when you're low on fuel or a storm is coming or there's something wrong with the plane. Making life or death calls, without hesitation, is a perishable skill; flying T-38s ensures we maintain it.

Even during an uneventful flight, it's crucial that you're focused and ready to work any problem that arises. When you're 150 feet off the ground and moving at 400 knots, which is common for fighter and test pilots, you have to concentrate on what's directly in front of you. If you don't, you'll die. That kind of intense focus is less about what you include than what you ignore. And by ignore, I mean completely block out; the argument with your boss, your financial worries—gone. If it doesn't matter for the next 30 seconds, then it doesn't exist. You need to be able to disregard everything that isn't going to happen in the next mile or so. There is only one essential question: What's the next thing that could kill me? Focusing on that thing, whatever it is, is how you stay alive.

Of course, luck has something to do with it, too. I once yanked back on the stick while practicing one-on-one fighting in a CF-18, and accidentally unplugged my g-suit. CF-18s have a heads-up display (HUD) that looks like a glowing, green projection in front of the windscreen; you never have to look around the cockpit, all the key information is there in that display. A video camera films the HUD, and afterward we always watched the HUD tape to see what had happened, so we could debrief. That's how I know I was unconscious for 16 seconds after my elbow hit the g-suit hose, unplugging it while the plane was pulling all the g it could pull, whereupon the blood promptly drained out of my head and I passed out. When I came to, I thought, "Wow, that was a really good sleep. I feel great. Hmm, that sounds like someone on the radio; maybe I'm still dreaming. Weird—the guy sounds a bit like Denis. Wait a sec. That is Denis. What's going on? Am I in an *airplane*?" All this was just rattling through my head until I opened my eyes and figured out that I was indeed in an airplane, and Denis, the other pilot, was practicing gunning me and wondering why I was flying so poorly.

Those 16 seconds were plenty of time to kill myself and him. Luck is what kept me alive while I was unconscious. Operational awareness—being able to see the big picture and focus on what could kill me next—is what kept me safe after I regained consciousness. I didn't waste a second thinking about why I'd passed out. In a crisis, the "why" is irrelevant. I needed to accept where I found myself and prioritize what mattered right that minute, which was getting back on the ground ASAP. There would be plenty of time later to try to figure out the why. (And we did: as a result, the g-suit connection in the CF-18 was altered so that it couldn't be unplugged accidentally.)

If you're focused on the wrong things, like the bee in your helmet or whose fault it is that the g-suit came unplugged, you are likely to miss the very narrow window of opportunity to correct a bad situation. In a real emergency in a fighter—an engine failure during takeoff, for example, or a fire in the cockpit—there's usually just a split second or two when the decisions you make will determine whether you live or die. There's no time to consult checklists. You need to know the boldface, the actions that are absolutely critical to survival—so called because in our training manuals, they're written in boldfaced capital letters.

"Boldface" is a pilot term, a magic word to describe the procedures that could, in a crisis, save your life. We say that "boldface is written in blood" because often it's created in response to an accident investigation. It highlights the series of steps that should have been taken to avoid a fatal crash, but weren't.

* * *

In Bagotville in 1986, my best friend, Tristan de Koninck, and I had one of the ultimate male bonding experiences: we went to the base clinic together to get vasectomies. This was a non-negotiable condition for remaining married to Helene after she became pregnant with our third child, Kristin; Tristan was the father of two little girls, about the same ages as our boys, and his wife believed their marriage would also benefit from a little less fertility.

At the clinic, I went in first and made a point of screaming and carrying on during the procedure, knowing that Tristan could hear everything in the waiting room and hoping I'd spook him. It didn't work; he had been a Snowbird, an aerobatic pilot, for two years before he began flying fighters, and had nerves of steel. We hobbled out of there sore but laughing.

About a month later, I was deployed to Bermuda. Tristan, back home in Canada, flew his CF-18 in an air show in Prince Edward Island. It was overcast, about 300 feet of cloud, when he took off the next day to fly back to Bagotville. He stayed low at first, then pulled up into the cloud. About a minute later he came straight back down at 700 miles an hour. The crash obliterated the plane; all they ever found of my friend was a little bit of his heel. It was inexplicable. Tristan was an excellent pilot, a much better formation flyer than I was.

I flew back to Bagotville for the funeral service, where I played his guitar and sang "This Old Guitar," which we used to perform together. It was one of the hardest things I've ever done. I had to practice singing it a hundred times, at least, until I was able to get through the song without breaking down.

Afterward, I worked on the accident investigation, but we were never able to figure out why Tristan's plane had crashed. He hadn't made a radio call, and the plane's telemetry and radar data were inconclusive. At the time the CF-18 was having various subtle failures of the displays that show you which way is up; perhaps that's what happened. There are also a lot of things that occur inside your body during a rapid acceleration; your own balance system may feed you inaccurate information. When you pull up hard into a cloud, you're relying on your instruments to tell you what to do, so if they lie to you, or if you're debilitated by dizziness, you might come screaming back down vertically without being aware of what is happening. Or there might have been another cause altogether. We'll never know. The only thing I'm certain of is that Tristan knew the boldface, and even that wasn't enough to save him.

Knowing the boldface improves your odds, but it's no guarantee. You can be the best driver in the world with the safest car in the world, but if a semi comes through a stop sign and plows into you, none of that will matter. Intellectually, I'd always understood that, but losing a close friend, one I knew to be at least as good a pilot as I was, really drove the point home. Looking on the dark side, sweating the small stuff, viewing your colleagues as the last people in the world, knowing the boldface and recognizing when to use it—in the end, none of it may save you. But in a real crisis, what other hope have you got? The more you know and the keener your sense of operational awareness, the better equipped you are to fight against a bad outcome, right to the very end.

Tristan was the first close pilot friend of mine who died doing his job; after that, however, I lost a pilot friend almost every year. It's a part of flying fighters, we all know that going in, but you never get used to it. Each loss is a sharp shock, followed by a wave of grief. I never felt that an airplane had snuffed out a friend's life, though; rather, a set of unusual circumstances was to blame. So the cumulative impact was not to make me afraid to fly, but to make me even more determined to understand what could be done to enable me and other pilots to work tough problems.

As a test pilot at Pax River, I spent years trying to figure out how to make fighters safer by thinking through, in a systematic way, what could kill a pilot—and coming up with new boldface to help prevent it. The goal was to give regular fleet pilots who'd just graduated from basic jet training everything they'd need to know to work a problem so that even if they couldn't save their planes, they could eject and save themselves.

We did this by putting F-18s out of control, deliberately, and figuring out how to get them back under control. It was a wild experience, physically, a little like getting on a roller coaster at the fair: it's comfortable enough chugging upward, but when you start whipping down there's a sense of rising panic and a feeling of unpredictable, external force. Amidst the violent accelerations and nauseating rolling and tumbling, you're responsible for keeping track of a lot of things, like your altitude and your engines, which may choke because of the changing intake air pressure. Meanwhile, you're also trying to quantify the experience: What's the rate of rotation? How hard do you have to grab the stick?

Working as a test pilot in the "out of control" program reinforced my ability to focus on the essentials even in the midst of chaos. I learned never to give up on a problem and never to assume that everything will turn out fine. It didn't occur to me, though, that the place where I'd really need to put those lessons into practice was on the ground.

If I hadn't understood how to focus and work a problem, I would not have got to space a third time. As it was, I just barely made it.

* * *

In 1990, when I was a test pilot at Pax River, I went back to Stag Island with my family for a holiday in late August. Shortly after we arrived, my parents threw a big party, the kind of event where people mill around the barbecue playing guitar and drinking whisky and eating their weight in corn and hot dogs. That night I woke up with gut pain. Any time I ate a lot I tended to pay for it, but this was different. I was in agony, and when the morning came I headed to Sarnia General Hospital. They put me on morphine, at which point I began hallucinating vividly about roller coasters and spiders, my dad became convinced that I was dying of cancer and the doctors began talking about exploratory surgery.

Alarmed, Helene got Charlie Monk, a physician and friend from Stag Island, involved. She explained to him that if I wasn't back at Pax River in a few weeks as scheduled, healthy and fit, I could lose my flight medical. A military pilot's career depends on medical clearance to fly; lose that, and you're toast. Abdominal surgery is particularly problematic: if you're in a fighter jet pulling g, the added load on your abdomen could rip the stitches open right there in the cockpit. Charlie explained this to the doctors who were treating me, but after three days, when I wasn't getting better and they still hadn't figured out what was wrong, they threw up their hands and decided surgery was the only option.

After opening me up, they did find the problem: a single strand of scar tissue, formed after my appendix was removed when I was 11 years old, had bridged onto my intestine and, like a drawstring, was pulling it closed. The surgeon snipped that strand and sewed me back up, leaving an impressive, jagged 8-inch scar across my belly. But I felt just fine. Two weeks later I was riding horses, and when we got back to Maryland, the U.S. Navy doctors checked me out and cleared me to fly. A month after being released from the hospital, I was back in an F-18.

At the time, it seemed like a little too close a call. But that medical emergency turned out to be a really lucky break. Had the constriction not been addressed it would likely have been discovered in 1992, during the astronaut application process, and I would have failed NASA's medical exam. A problem I wouldn't have even known I had could have finished my chances of becoming an astronaut. Applicants are regularly ruled out for more minor medical conditions.

Over the next two decades, my most serious health problem was a head cold. I passed the physicals for my Shuttle flights, no problem, and in 2001 I passed the most stringent medical exam in the world and was certified to go on the ISS. Then in the late fall of 2009, the crew for Expedition 35 was selected and I was told I'd be commander. It was something I'd been working toward my whole adult life, and I was both proud to get the assignment and humbled by it. I wanted to be worthy of the honor, to vindicate NASA's trust in me and the CSA's investment in me—it was the first time a Canadian would command the ISS, and only the second time that the position had been assigned to an astronaut who wasn't American or Russian.

A crew is trained to look after everything on board, from the potable water dispenser to all systems in the Japanese module, but there are varying degrees of expertise. Being certified as a user means you have basic knowledge and can turn things on and off; operators can run a module or system unaided, and know how it works but not how to fix it; specialists can do it all—operate, understand and repair. Becoming a specialist in all modules and systems would require considerably more travel and hundreds of hours of extra training, most of which I would in all likelihood never need to put into practice on the ISS. But that was all right. I decided to try to be designated a specialist in as many modules and systems as possible. This was my last chance to make a real contribution to the space program, since I would never get another opportunity to leave Earth.

By October 2011, I was a specialist in almost every ISS system, experiment and module. I'd been training hard for two years, regularly working nights and weekends, and spending 70 percent of my time either in Russia or elsewhere on the road. I was happy to be back in Houston with Helene for a few weeks, only my stomach didn't feel quite right. She was recovering from the flu, so I figured I'd caught it too but decided to go to the NASA clinic, just in case. The doctor there didn't think my problem was the flu. He sent me to the hospital, suspecting an intestinal obstruction. An MRI confirmed it.

This was not good news, but sometimes a blockage will clear on its own. That's what I hoped would happen, but it was one of those hospital stays where everything that could go wrong, did: they accidentally wound up dehydrating me, and then after three days, when I was much sicker than I had been when I was admitted, the surgeon assigned to my case announced that he'd be operating on me the next day. He wanted to do what the surgeon in Sarnia had done back in 1990: make a big incision in my abdomen, open me up and see what the problem was. In the intervening two decades, however, laparoscopic procedures had become much more common; these involve a tiny incision and the use of a laparoscope to transmit images to a video monitor. Because laparoscopic surgery is minimally invasive, there's a much lower risk of complications than with traditional surgery, and recovery time is also minimized.

Given what I'd just been through there, the prospect of having an operation at that hospital didn't appeal to me. Furthermore, I knew that if the surgeon operated the conventional way, with scalpel and large incision, I would not be going to the ISS in 2012. I would be medically disqualified. But I might still have a shot if I could get a laparoscopic procedure—and if it turned out that the issue was in fact minor. We had 24 hours to work the problem and I was, by this point, really feeling ill. Helene got on the phone and in short order I was moved to another hospital where I received excellent care. I was soon scheduled for laparoscopic surgery with Dr. Patrick Reardon, who'd treated Barbara Bush.

He made two very small incisions in my abdomen and, using flexible snake-like devices just 3 millimeters wide, quickly located the problem: the surgery back in 1990 had created a 1.6-inch adhesion—a glob of sticky scar tissue, basically. The vast majority of abdominal operations result in adhesions, and adhesions are in turn one of the most common causes of obstructions because they can twist or pinch the intestines closed. That was exactly what was going on here: this adhesion, likely inflamed by the flu virus, was essentially gluing my intestines to my abdominal wall. When Dr. Reardon released the adhesion, everything sprang back into its proper place. After carefully inspecting my insides, he closed me back up and told me I should have no further trouble.

I knew this wasn't accurate, though. Now there was a whole new problem to work: convincing the powers-that-be that I was healthy enough to go to space. On the plus side, I didn't have a chronic condition and one of the top surgeons in North America thought I was good to go. However, if I had a recurrence in space, our mission would be cut short and we'd have to fly home early. Another crew would have to launch earlier than planned to replace us. The cost would be astronomical.

Before I could persuade anyone else I was fit to fly, I first had to convince myself. I wanted to go to space again, of course, but if there was any chance of getting so ill that I'd need to be evacuated from the ISS, I had a responsibility to withdraw from the expedition. I needed to find out what the risk of a recurrence really was, so Helene and I started researching and talking to doctors. In the meantime, I felt perfectly fine and was cleared to go back to training—but I wasn't cleared for space flight. Every country that funds the ISS would have to sign off on that, which would be a tall order given the stakes.

Over the next two months, a panel of experts—surgeons, military doctors, authorities on the medical aspects of space flight—considered the issue in order to make a recommendation to the Multilateral Space Medicine Board (MSMB), which includes representatives from the U.S., Canada, Europe, Japan and Russia. In order to decide whether I was a good statistical risk or not, they needed statistics. So a medical doctor was hired to review the research on the likelihood of another obstruction after surgery. But as it turned out, most of the studies had been conducted before laparoscopic surgery was common; many of them lumped together people who'd had minor procedures like mine with people who'd had really serious problems like massive internal trauma after car accidents or operations to remove tumors. And these studies did show that the risk of a future problem was unacceptably high: 75 percent.

I'm no medical expert, but common sense told me that that data had little bearing on my situation. My problem had been minor, and it had been repaired using the latest and least invasive technology. Dr. Reardon had told the MSMB that the risk of me having another intestinal obstruction while I was on Station as just one-tenth of one percent. The chances that we'd have to evacuate the ISS to get me home were, in other words, significantly lower than the chances that an astronaut would have to be evacuated for a tooth abscess.

I felt it was important to put even that very minimal risk in context; going to space is inherently dangerous, and activities such as spacewalks compound the danger. Seen in that light, the risk of a recurrence was inconsequential. I made my case directly to the two Canadians who served on the MSMB, presenting as much information about laparoscopic procedures as I could so that they were well prepared for the meeting. When the members of that international panel convened in November 2011, their ruling was unanimous: they cleared me for space flight, though they wanted to see some of Dr. Reardon's studies.

Phew. All's well that ends well. Only, it wasn't really over. Two months later, I learned that some doctors at NASA hadn't been satisfied that I really would be all right, and had gone to their Canadian counterparts asking for more proof—but like a lot of top doctors who are in demand, Dr. Reardon hadn't had time to publish his results. He didn't have a neatly printed academic journal article to show them, just his own expert opinion based on extensive experience. So, unbeknownst to me, a new panel of four laparoscopic surgeons had been asked to consider whether it would be a good idea to have what they kept calling "a quick look inside"—in other words, to perform exploratory surgery to see whether I really was okay or not.

No one had breathed a word of any of this to me or to the flight surgeons at NASA who would be directly responsible for my health while I was on the ISS. The secrecy and paternalism really bothered me. They trusted me at the helm of the world's spaceship, but had been making decisions about my body as though I were a lab rat who didn't merit consultation. One thing I'd learned was that I couldn't expect every medical professional to be an expert on every single medical problem and procedure. The information we'd unearthed on our own had been crucial so far, and so had the opportunity to frame the medical risks in the context of the overall risks of space flight. Keeping me out of the loop only made sense if the experts were omniscient and I had nothing to contribute to the discussion.

The reasoning also bothered me. Just as a panel of hairdressers is likely to recommend that you change your hairstyle, a panel of surgeons is likely to recommend surgery. And that's exactly what happened, even though three out of the four surgeons thought the chance of a recurrence was low or nonexistent.

So in January, I was asked to have surgery yet again. My starting position—"I will do it, but only if you absolutely insist on it"—quickly changed to a firm "no." Helene and I had been researching like crazy, and the more we learned, the more this "quick look" idea seemed truly idiotic. There were, it turned out, two studies covering cases exactly like mine: after conventional surgery people had developed obstructions, which were then cleared laparoscopically. The rate of recurrence? Zero. To me, this was the best proof possible that I was a good risk to go to the ISS, particularly since in all other respects, my health was and always had been excellent. Plus, like any surgery, the procedure that was being proposed would introduce significant new risks. I might have an adverse reaction to general anesthetic, for instance, or I might develop an infection, or any number of surgical errors might occur—and any one of those things could then eliminate me from space flight. A needless operation simply made no sense for me and would also establish a troubling precedent. What about the other 20 percent of astronauts who'd had appendectomies and therefore might also have adhesions? Would they too be required to have "quick look" exploratory surgery?

There was something else to consider: the risks to the space program itself if I didn't fly. I was backup for another commander, Sunita Williams, who was scheduled to launch in July.

Who else could step in to cover her? The answer at that late date was, “No one.” Like Suni, I was left-seat qualified for a brand-new spaceship, the Soyuz 700 series, which is digital rather than analogue and therefore has different flight control displays and laws. If I was pulled, the CSA couldn't replace me; no other Canadian was even qualified to fly the older type of Soyuz, let alone the new one. NASA couldn't replace me either: my NASA backup was an astronaut who'd never been to space before. He was very competent, but he couldn't possibly get qualified by July. In order to swap in someone who was qualified, as the Chief Astronaut at JSC pointed out, five crews would be affected, so there would be a significant ongoing safety risk to the entire program. If the likelihood had been high that I'd get seriously ill in space, there would have been no option but to take those risks. But the chances weren't high. In fact, they were so low as to be negligible.

The next few months of my life, while I continued to train and to get ready for an expedition I might or might not lead, were Kafkaesque. I had to try to focus on training and on learning everything I could, and ignore the background noise. I was caught in a bureaucratic quagmire where logic and data simply didn't count; internal politics and uninformed opinions were what mattered. Doctors who hadn't ever performed a laparoscopic procedure were weighing in; people were making decisions about medical risks as though far greater risks to the space program itself were irrelevant. Helene and I, along with our flight surgeons, were spending vast amounts of time and energy digging up studies, talking to experts, emailing administrators, creating complicated graphs and charts comparing medical data and different risk factors—just looking for some other way to persuade administrators that it was safe for me to fly.

Meanwhile, the MSMB ruled: all the evidence we'd dug up convinced the international members on the board that I was fine to fly, but not the American, who wanted still more proof.

This was not good news. We'd presented what we and the experts who'd helped us considered to be overwhelming evidence. If that didn't do the trick, what would? It felt as though we were mounting a case against superstition, which science is useless to dispel. You can present all the random sample studies you want to prove that it's safe to walk under a ladder, but a superstitious person will still avoid that ladder.

The CSA kept telling me to relax and not worry; they were sure that in the end everything would go our way. This was completely in keeping with national character: Canadians are famously polite. We're a nation of door-holders and thank you-sayers, but we joke about it, too. How do you get 30 drunk Canadians out of a pool? You say, "Please get out of the pool." Under normal circumstances, Helene and I would be the first people out of that proverbial pool, but these were not normal circumstances; we felt the Canadians were being just a little too Canadian, trusting that logic would eventually conquer all. To us, it was plain as day that our data collection efforts were crucial, both for me personally and to protect Canadian interests. Many millions of dollars had been invested in my flight; many Canadian experiments were slated to go on board during my expedition, too. Having a Canadian in command of the world's spaceship was not only a source of patriotic pride but also a vindication of the space program, whose funding, like NASA's, is perennially under threat. If we stopped working the problem, I wouldn't be going to space, though, since a single individual in senior medical management at NASA could prevent it.

And then, at the eleventh hour, just days before a March

meeting where NASA would decide once and for all, someone on the MSMB volunteered a solution: an ultrasound would likely reveal whether I had another adhesion. I was dumbfounded. For months I'd been asking whether there wasn't another way, something less invasive than surgery, and for months the answer had been, "No, surgery is the only possible option." Now, suddenly, everyone was on board for an ultrasound, so long as one particular highly qualified radiologist performed it—he was, however, on holiday, so I had a week to do some research, long enough to discover that the ultrasound test had a 25 percent rate of false positives. In other words, the test might determine that I did have an adhesion, but there was a one in four chance it would be dead wrong. And even if I did have an adhesion, who was to know whether it would be threatening or not? No one seemed concerned about this but me and Helene.

When the day came to get the ultrasound, we were both resigned during the 45-minute drive to the hospital. We had fought the good fight right to the very end. Now it was time for a death sim of sorts: we needed to talk about what we'd do when I failed the ultrasound. We discussed a lot of different options: staying in Houston longer than I'd planned, maybe, or retiring and looking for work as an aerospace consultant.

The main thing we decided during that drive is that we would not be defined by this experience. I wouldn't go through the rest of my life being the commander-who-wasn't, that poor guy who didn't get to go to space a third time. We'd seen what had happened to other astronauts who were scrubbed from missions, and we thought that the next thing that would kill us, metaphorically speaking, wasn't an ultrasound but a loss of our own sense of purpose. Fortunately, we also knew the boldface that could save us: focus on the journey, not on arriving at a certain destination. Keep looking to the future, not mourning the past.

We arrived at the hospital feeling pretty good. Whatever happened, we knew we would be all right. The expert plunked goop on my stomach, then used different ultrasound wands to look at the area. The inspection didn't start off well. The doctor said, "Oh, that wasn't what I expected to see"—he needed to observe movement, what's called "visceral slide." He turned the monitor so that I could watch, too. Helene was holding my hand, her back to the screen, tense but resigned. A minute passed. Even I had to admit that nothing was moving.

I'd failed. But more than disappointed, I felt curious: Had I really been so wrong? Was there something wrong with me after all? So, my eyes glued to the monitor, I started breathing more shallowly, tensing and relaxing my stomach muscles, actively willing my insides to slide. I wanted to go to space, of course, but I also wanted to be certain that I actually was okay.

After years of studying and training, this was what it all came down to: whether a minuscule portion of my intestine could move on command. And then, miraculously, it did. The doctor smiled, and turned on a recorder to capture the movement on video: visceral slide. Another doctor came in and verified it, and then the relief in the room was palpable.

Back in the car, Helene and I started calling the few people who'd known about this whole ordeal. We felt we'd won an epic David and Goliath sort of battle, one I'd been getting ready for, without knowing it, my entire adult life. It had been the ultimate "out of control" test, working a serious, complex problem while in freefall, professionally, without losing my focus on the true goal of the mission: making sure our crew was ready for space flight whether I was going with them or not. But there wasn't time to celebrate the victory. I had work to do.

I was going to space, after all.

PART II

LIFTOFF

7
TRANQUILITY BASE, KAZAKHSTAN

A LOT OF PEOPLE ASSUME that the days right before launch are some of the most stressful ones in an astronaut's life. Actually, the opposite is true: the week or so pre-launch might be the closest we ever get to serenity, professionally speaking. One reason is that nothing has been left to the last minute. We've been preparing for this specific launch for years, and thinking and dreaming about space flight most of our lives. The other reason is that we're in pre-flight quarantine. Astronauts call it "white-collar prison," only half-jokingly; we have minders, we can't leave the compound and most visitors have to talk to us through glass. But of course we *want* to be there, and we're catered to, fussed over and waited on so attentively that the casual observer might never guess the true purpose of our stay is medical. The idea is to protect us from catching infections on Earth that would make us sick—and less productive—in space.

On orbit, even a head cold is a big deal. Without gravity, your sinuses don't clear and your immune system doesn't fight back as effectively, so you feel much sicker, much longer—and in such a confined space, it's pretty much guaranteed that the rest of the crew will be infected. That's exactly what happened during the

Apollo 7 mission in 1968. Commander Wally Schirra developed a bad cold partway through the 11-day mission, and by the end, all three members of the crew were so ill that they refused to put their helmets on for landing. They were concerned that as pressure increased during re-entry, their eardrums might burst, so they wanted to try to equalize the pressure the same way you would on a plane: by pinching their noses while trying to blow out—which would be impossible if they were wearing big fishbowl helmets. The crew's exchanges with Mission Control in Houston were famously fractious, and none of those three astronauts ever flew again. In later years Schirra did, however, appear in ads for Actifed, the decongestant he'd taken in space.

In the 1960s, astronauts frequently launched in apparently perfect health, but then, a day or so into the mission, a virus would make its presence known. The crew of Apollo 12 also wound up relying on Actifed; all three astronauts on the Apollo 8 mission experienced gastroenteritis, which is probably even less pleasant on orbit than it is on Earth. But not until 1970 did NASA decide it might be a good idea to isolate crews pre-flight. Apollo 13 was the last straw: three days before launch, a backup was swapped in to replace a crew member who'd been exposed to measles (but didn't, as it turned out, ever fall ill). In flight, in the midst of a life-threatening crisis—an oxygen tank had exploded, causing serious damage to one of the rocket's modules—another crew member came down with an infection. Thereafter, pre-flight quarantine became mandatory.

When the Shuttle was still flying we spent six or seven days in quarantine, roughly the length of time it would take a virus to run its course. At the Kennedy Space Center (KSC) our quarters were spartan—stark little rooms with a dresser and a hard bed, like a military barracks—but the mood was convivial. For crew members, launch was a momentous event, of course, but the Shuttle left Earth with seven people on board on a fairly regular basis. The people at KSC were accustomed to sending astronauts off to space. By the time the program ended in 2011, there had been 135 launches, most of which never even made the nightly news.

These days, when the Soyuz is the only manned vehicle going to the International Space Station, and it departs not from sunny southern Florida but the near-desert of the Kazakh Steppe, the whole experience of quarantine is different. Now just a dozen humans leave the planet each year and we stay in space for months rather than a week or two—long enough to start feeling at home there, and long enough that anything could happen in our absence. The knowledge that something bad might happen to people we love while we're in space, yet there would be nothing we could do to help and no way to come back early, somehow injects a slightly more formal and contemplative flavor to the whole experience of quarantine.

Another difference: the Russians, so austere and no-nonsense in their approach to many things, are big believers in downtime for space explorers. We're quarantined longer in Baikonur, Kazakhstan, than we were at the Cape—12 full days—but you get the sense that Roscosmos doesn't think that's quite long enough. Before my last flight, my crewmate Roman was sent with his family to a health retreat in the country for five days *before* quarantine, to kick-start the unwinding process. (Post-flight, too, cosmonauts get months off work, while astronauts go back to the office just a few weeks after returning to Earth, though we're certainly not expected to take on a full slate of responsibilities the moment we walk through the door.)

These days, the purpose of quarantine is as much psychological as it is medical: an enforced time-out ensures we pause, consider what we are about to do and deliberately begin to transition to a new kind of existence. Emotionally and physically, quarantine is a half-way house en route to life in space.

* * *

As we left Star City, Russia, to fly to Kazakhstan in December 2012, it was the usual mad scramble to get to the plane, then . . . calm. Tom, Roman and I were heading out of the murk of preparation and into the clarity of launch: the plane was almost palpably full of plans, hopes and dreams. Looking out the window as we made our descent, however, I found the view less than enchanting. The Syr Darya river flowed darkly through the flat brownness of the landscape, which was otherwise punctuated only by a scattering of low-slung, utilitarian apartment buildings festooned with satellite dishes. There were no hills and very few trees. It looked exactly like the kind of place where a rocket could crash without inconveniencing anyone or even, possibly, attracting notice.

Baikonur is a spaceport: space flight is its main industry, its reason for being, yet there's nothing remotely slick or futuristic about the place. Nor is it hopping—seasonal extremes of temperature don't encourage colorful street life. In summertime, the heat is oppressive, but when we arrived in December, it was so cold that after a few minutes outside under a bright blue sky, frost formed on the tips of my eyelashes. On the outskirts of Baikonur, camels wandered in and out of holes in the fences while stray dogs howled at the approach of winter. It felt like a ghost town redolent of history and Soviet-era, matter-of-fact triumphalism. The tree planted by Yuri Gagarin, the first human being to leave the planet, was somehow thriving on an otherwise barren plain.

There was an "as if" quality to my first day there, partly because the town itself is a little bizarre, both otherworldly and prosaic, but partly because I'd come so close to not being there at all. Checking into what's known affectionately as the Cosmonaut Hotel helped me believe that yes, this was really happening. While the tourist attractions of Baikonur may not rival those of Cape Canaveral, it must be said that crew quarters are considerably more spacious: I had a whole suite of rooms, complete with a massive Jacuzzi. The overall ambience brings to mind the institutional charm of a sizable college dorm. Astronauts and cosmonauts are housed in one wing, while another includes support staff and instructors; there are Ping-Pong and billiards tables, a serious gym and a dining hall. Everything is spotless (sterile, actually: the floors and walls are wiped down with bleach daily to ward off germs) and the food is great (the kitchen staff is fanatical about hygiene, so there's no chance we'll get food poisoning).

Breakfast is oatmeal, yogurt, *tvorog* (Russian cottage cheese), omelet with red caviar, persimmons and honey, nuts and fruit compote, and to drink there's coffee, tea or chicory. Lunch and dinner are varied banquets of homemade soup, grilled fish, cutlets, *pelmeni* (the Russian version of ravioli) or *manti* (a Turkish dumpling stuffed with meat), fresh vegetables and made-to-order dessert. Ask for a brownie, and they will cheerfully whip up a fresh batch, packed with nuts and topped with chocolate sauce.

Early in the morning of our first full day in Baikonur, we finally got to see our Soyuz—the real one that would actually take us to space. Back in the summer, we'd met with a delegation of the rocket's builders for a traditional toast to success and friendship—with a little symbolic sip of rocket fuel, which, even when it's cut with water, tastes like kerosene: just awful. In subtle ways, the rocket they built for us was different from the simulator in Star City; after almost every mission the vehicle's design is tweaked a bit. During the *primerka*, or fit check, we spent about an hour inside, all suited up in our Sokhol pressure suits, figuring out where the switches really were and how long it would actually take us to do things. We were satisfied: the ship was sturdy and familiar.

The rest of our days in quarantine were full but tranquil. We did not-so-taxing things like packing the personal items we wanted to take to space. That didn't take long, because the Soyuz is so small that weight and balance affect how it flies; the designated bag is the size of a small shaving kit. I managed to cram in a new wedding ring for Helene, some commemorative jewelry, a watch for my daughter, Kristin (I flew a watch each for my two sons on previous flights), a full family photo for my mom and dad, and some guitar picks emblazoned with our Expedition 35 emblem—all things I could later give to people as "flown" gifts.

In quarantine we also worked out, though carefully, particularly after one Russian manager blew out his Achilles tendon playing indoor badminton. I knew that if that happened to me, I'd be headed for Houston, not the ISS. At this point, my departure wouldn't be such a huge problem for NASA, because my replacement was just down the hall at the Cosmonaut Hotel; the backup crew does everything the prime crew does, right up to the final hours before launch. To guarantee that the show could go on even if disaster struck, the two crews travel to Baikonur in separate vehicles. Just in case.

Training never stops in our business, not even when we're on the ISS, but it does slow down considerably in the days leading up to launch. We'd already been deemed fully competent to fly—we had completed "final quals" and signed the traditional pre-launch book in Yuri Gagarin's old office in November. So in Baikonur we just took some refresher classes: reviewing the lessons learned during recent missions, for example, and practicing docking the Soyuz on a portable simulator. Overall, the workload was light and included things like media interviews (at a safe, germ-unfriendly distance). We also signed endless stacks of crew photos, enough for each citizen of Russia, it seemed.

While the backup crew toured local museums (cautiously, treating other people as walking disease vectors), we remained cloistered, reading books and taking advantage of the Wi-Fi (on the ISS, the Internet is dial-up-era slow). In the evenings we'd reunite and, along with our instructors and support staff, head to the *banya*, the Russian version of a sauna. Afterward, we often played guitar and sipped single malt, a group of friends from all over the world, united by our mission.

Everyday routines and stressors had been stripped away to encourage us to focus—emotionally, intellectually and physically—on our mission. At first I felt a little unmoored: after years of studying and rehearsing, suddenly there were very few formal demands on us and no difficult challenges to face. But pretty quickly, I adjusted to a simplified existence. Freed from everyday responsibilities such as making my own meals and doing my own laundry—as all astronauts do at least in Star City and on the road, if not in their own homes—I took it easy and had a chance to gather my thoughts. Tom, Roman and I were about to go away for quite a few months and take quite a few big risks. The best thing we could do for ourselves was to let that reality dominate our mental landscape until seriousness of purpose met buoyant certainty: yes, we're ready to do this thing.

As our time in quarantine drew to a close, I felt more confident and focused every day. I doubt I would have had the same sense of readiness if someone had told me, "Okay, show up in Baikonur on Wednesday morning, you're going to space at noon." I'd probably have spent the previous day running around doing all the things everyone does before a trip: packing, paying bills, picking up dry cleaning. Even if you're highly competent, when you're careening full-speed toward a deadline or a destination, you usually arrive breathless, still mentally scanning your to-do list and not fully focused on the task ahead. You may achieve impressive results anyway, but you're likely to deliver less than you would if you didn't feel harried. For me, anyway, going into a high-pressure situation feeling calm and fully prepared has another benefit, too: I'm able to live more fully in the moment, absorbed and engaged in it, and better able to appreciate it as it unfolds rather than in retrospect.

* * *

Of course, that kind of single-mindedness takes a village—other people have to pick up the slack when you're unavailable, literally or figuratively. If you fail to recognize that fact and behave accordingly, you can count on creating exactly the kinds of distractions and conflicts you should be trying to avoid when you're facing a major challenge. People around you will let you know in no uncertain terms that your single-minded dedication bears a striking resemblance to pigheaded selfishness.

During our first few years in Houston, I'd volunteer for anything and everything at NASA and the CSA, so I was on the road a lot. After a while I started to notice that when I got home, there was no longer a hero's welcome. The kids didn't leap up and rush joyfully to the door to greet me. Sometimes they even seemed a tad annoyed to see me, particularly if I reminded them of my expectations in the way of manners, rules and comportment. Helene was delighted to explain this puzzling phenomenon. She informed me in the most diplomatic fashion possible that I'd been away so much that my family had learned to live without me, and she and the kids had developed their own ways of doing things and didn't really appreciate my attempts to turn back the clock. In other words, I was now effectively a visitor in my own home and would have to put in some serious time before picking up the threads of fatherhood. She went on to say that she'd wondered if maybe I wasn't going just a wee bit overboard with the extra work assignments. Were they really getting me closer to my professional goals? Or had I simply got in the habit of saying yes at work and no to my family?

We'd had a similar discussion back in Bagotville, when we had three kids under the age of 5 and I was spending quite a few of my days off taking part in optional military exercises. Helene had asked, point blank, "Do you want to have a family or just a career? I'll happily give you the space to have both, and I'm willing to carry 90 percent of the burden here at home until I get a paying job again, but I can't carry 99 percent." She was all for me volunteering—but she encouraged me in the strongest possible terms to start evaluating on a case-by-case basis whether a given volunteer opportunity was something I needed to do for professional growth, or just something I wanted to do. I did try to prioritize differently after that and to be more conscious of the effects my decisions had on her and on my own relationship with our kids.

I had to recalibrate again in Houston. The reality of an astronaut's life is that you travel 70 percent of the time and you don't have much say over your own schedule—so when you do have leeway, you have to make choices that clearly communicate gratitude to your family and a desire to see them, on their terms, every once in a while.

In quarantine, however, there's no pretense of trying to balance work and personal life—your domestic responsibilities go right out the window and family life is pushed to the margin. That's the whole point. In Baikonur my family and Tom Marshburn's arrived along with a CSA/NASA contingent three days before launch, and stayed in a hotel just a stone's throw away from the crew quarters. We were allowed visits from our spouses and children, but only during strictly scheduled and relatively brief time slots and only after they had been checked by a doctor (even so, we were encouraged to keep them at arm's length). Extensive negotiations were required to get my brother Dave into crew quarters for 30 minutes so we could play guitar together and record a song—sitting clear across the room from one another, to be on the safe side. Tom's daughter, Grace, who was then 10 years old, didn't even get to be in the same room with her dad. Kids under 12 are considered too infection-prone and rambunctious for the monastic environment of quarantine, and can only interact with quarantined astronauts via phone, behind soundproof glass.

Although quarantine is designed to protect astronauts, it's certainly not painless for our families. For starters, they have to come to us, and Kazakhstan is not easy to get to unless you live in Kyrgyzstan. Then, not only are they at the mercy of our schedule, but they are required to take part in "fun" traditions that may not strike them as especially entertaining. A day or two before launch, for instance, we always watch *White Sun of the Desert*, a Russian movie with a *Lawrence of Arabia*–esque hero, with our crewmates and relatives (who may be considerably less amused than we are by the overacting).

For those of us who are going to space, rituals like this impose a reassuring, predictable structure on the days leading up to launch. For our families, though, these rituals may feel more like additional obligations when they're already carrying an extra load. Not only have we shrugged off all domestic duties, but our spouses are responsible for hosting the friends and relatives who've come to see us off. By the time we head to the launch pad, serenely focused on our mission, our spouses tend to be feeling pretty stressed. As my colleague Mike Fossum says, "Let's face it—our dreams become their nightmare."

It was even more stressful when the Shuttle was still flying. For my first space flight in 1995, Helene and I invited just about everyone we knew, along with everyone they knew, and wound up with more than 700 guests. *Hey, a Florida holiday that includes a rocket launch and a VIP badge from* NASA? *Sold!* About a week before the big day a horde of family and friends descended eagerly on Cocoa Beach, Florida. Even the name evoked a holiday feeling and they had a great time, golfing and going to Disney World, frolicking on the beach and painting the town, while their astronaut friend/relative was in lock-up. Of course, we wanted all of them to have a great time, but my role in ensuring that was pretty much limited to not dying. Helene, on the other hand, arranged a party, hosted endless breakfasts, luncheons, dinners and other events, and gave media interview after media interview ("Yes, I'm so proud!"). Non-stop mingling was the order of the day; people were understandably celebratory and in the mood to socialize, and they all wanted as much access as possible to the immediate family. She was basically run off her feet.

The launch from Baikonur in December 2012 was a slightly different story. I was allowed just 15 guests total, including immediate family, and it was wildly expensive to get there, via Russia, right before Christmas. Our closest friends and family, plus Tom's closest friends and family, plus a cadre of CSA and NASA people took over a hotel in Moscow. Helene and Tom's wife, Ann, helped arrange walking tours, provided restaurant recommendations and answered countless questions about what to wear, how to get to the subway station and when the bus to the airport was leaving. Helene told me it was like hosting a destination wedding. The only thing missing was the groom.

When after a few days the party relocated to Kazakhstan on an ancient plane chartered by NASA, the mood became even more festive. Jet lag, frigid temperatures that shocked even Canadians and a complete absence of language skills were apparently remedied with wild nights in various Baikonur "hot spots." When Helene and the kids trooped over from the hotel to see me for the hour or two we were allotted to be together each day, they brought increasingly colorful stories about sensible, hard-working relatives and friends who had, the night before, morphed into vodka-loving party animals with a taste for wearing other people's bras draped on their heads like berets.

Everyone had great fun, including Helene, but for her there was also the stress that comes with managing the logistics of a week-long reunion while worrying that something might happen to delay the launch. She was not, however, worrying about me, not even when we reviewed my will. She was counting on me to sweat the small stuff during launch and afterward, too. Also, she's a realist: she knows that exploration is risky, some explorers will die, and worrying won't change that fact. Some spouses are nervous to the point of nausea before launch, but mine was increasingly excited the closer we got to liftoff, and not just because my dream was coming true. There was pride and joy that I'd made it, but relief, too. She was ready to get back to her real life and her own adventures.

* * *

Fortunately, some smart person at NASA recognized long ago how difficult launch is for spouses and came up with the idea of family escorts: you choose two astronauts who aren't currently training for a mission, one to look after the immediate family and one who's in charge of extended family and friends at launch. Essentially, the family escort is a surrogate spouse: someone who's available to help out on Earth not only during launch but later, when life has returned to normal but the mission is ongoing. I've been a family escort a bunch of times, and the job includes running back to the hotel to get the access badge someone forgot, carting home the uncle who got bombed at the party, grabbing sandwiches, counting heads on the tour bus, dealing with complaints about hotel rooms that are too hot or too cold—you're essentially a dogsbody, but that never bothered me, not least because I knew I'd need someone else to do this stuff for my family if I ever went to space again. In 2012, that was Jeremy Hansen, a decorated fighter pilot and Canadian astronaut, who spent the days before Christmas herding my guests onto buses and in and out of museums, hauling their luggage from one place to the next, helping them exchange money and making sure they woke at 4:00 a.m. to catch their flights home—all with seasonal good cheer.

When you're choosing your family escort, you don't just consider which astronaut is most likely to be able to smile and nod when Aunt Ruby gets going on one of her political rants. Mostly, you think about which astronaut you'd want standing next to your spouse if someone you loved died while you were in space—or if your own rocket blew up, in which case the family escort would need to stand there for months or even years. For my second launch in 2001, Rick Husband was one of my escorts, and he did a lot of helpful things for my family. Next time he flew, his own family escorts, who included CSA astronaut Steve MacLean, had to step in and support his wife through the hardest experience imaginable: Rick was the commander of *Columbia*, the Shuttle that disintegrated on re-entry. Agreeing to be an escort, you know that you may wind up helping a spouse not only during a rowdy launch party but at a funeral and long afterwards, doing things like helping set up educational trust funds for the kids and advocating for the family during the accident investigation. I've never had to do any of that, thankfully, but you know it's a possibility when you agree to be a family escort. It's a huge responsibility.

But it's one we should take on, not just for altruistic reasons but for self-interested ones, too. Taking guests' orders for a Starbucks run and making sure someone else's grandpa has his preferred brand of gluten-free bread is a highly effective ego check. And there's something else: being an escort forced me to see the world through the eyes of the family of an astronaut. My own family had let me know on one or two occasions that being the child or spouse of an astronaut isn't always easy. Kristin puts it this way: "When your dad is an astronaut, the most interesting thing about you, growing up, doesn't have much to do with you, and it's nothing you control or influence. The fact that your dad is an astronaut trumps everything else people see when they look at you." My children dealt with and overcame this challenge in different ways; all three are now accomplished adults with full lives and many interests. But my career choice made that more difficult in some respects, and being a family escort helped me understand that many of the difficulties were situational rather than specific to our family. Helping colleagues' families during a launch, you become keenly aware of the ways that all families are forced to juggle and sacrifice—not just while their dad or mom or spouse is in space, but for years beforehand.

From 2007 onward, I spent about six months a year in Star City and also trained in the U.S., Japan, Germany, Canada and Kazakhstan. I was only home about 15 weeks a year and I missed a lot of birthdays and holidays. Inevitably my schedule created hardships for everyone who's close to me. There was no way around that, but I did try to anticipate potentially negative consequences so I could figure out how to prevent them. Long before

Little did I know that this early training in 1964 was actually getting me ready for flying in the tight quarters of the Soyuz spaceship. *(Credit: Chris Hadfield)*

My first flight suit—a proud young Royal Canadian Air Cadet off to glider pilot training, summer 1975. *(Credit: Chris Hadfield)*

Being awarded a scholarship to learn to fly gliders, spring 1975. My first step toward being a pilot. *(Credit: Chris Hadfield)*

Helene and I happily marrying in Waterloo, Ontario, on December 23, 1981. I was still a military college cadet, so wore my formal scarlets. *(Credit: Chris Hadfield)*

Test Pilot School, Edwards Air Force Base, California, as one of my toughest—and most fun—years of training came to an end in December 1988. A big day for the family, station wagon loaded to depart for Patuxent River, Maryland. *(Credit: Chris Hadfield)*

The whole family—Helene, me, Evan, Kyle and Kristin—together for Christmas 2005, at our home near the Johnson Space Center. *(Credit: Chris Hadfield)*

Flying a U.S. Navy F/A-18 with a hydrogen-burning research engine on the wingtip, being chased by a NASA Dryden 2-seater, at Pax River, 1991. *(Credit: Chris Hadfield)*

Dragging supplies through the snow with American astronauts at Winter Survival with the Canadian Army in Valcartier, Quebec, February 2004. *(Credit: Chris Hadfield)*

Mission Control in Houston, Texas, CAPCOMing for Space Shuttle mission STS-77 in 1996. My kids hand-painted my tie for Father's Day. *(Credit: NASA)*

Checking my gloves, ready to spacewalk: a day's training underwater in the Neutral Buoyancy Lab in Houston, Texas, 2011. *(Credit: NASA)*

Out in the untrespassed sanctity of space, between the Earth and forever. Canada's first spacewalk (and mine!), April 2001. *(Credit: NASA)*

Signing the traditional pre-launch book in Yuri Gagarin's office with Roman and Tom and our backup crew in Star City, Russia, November 2012. *(Credit: NASA)*

Inside the International Space Station Cupola, able to look down onto the whole world. A marvelous place to play guitar. *(Credit: NASA)*

The crew of Expedition 34, cool in sunglasses aboard the International Space Station. Someone had said, "Okay, a serious picture!" *(Credit: NASA)*

Dressed for spaceflight! Back in my Sokhol pressure suit, ready to return to Earth in our Soyuz capsule after five months on the International Space Station. *(Credit: NASA)*

Tom, Roman and me in our Soyuz spaceship, designed and trusted to be small and rugged enough to safely deliver us down through the fiery atmosphere, home to Earth. *(Credit: NASA)*

The International Space Station. *(Credit: NASA)*

The Soyuz undocking from the International Space Station, with Tom, Roman and me inside, May 13, 2013. *(Credit: NASA)*

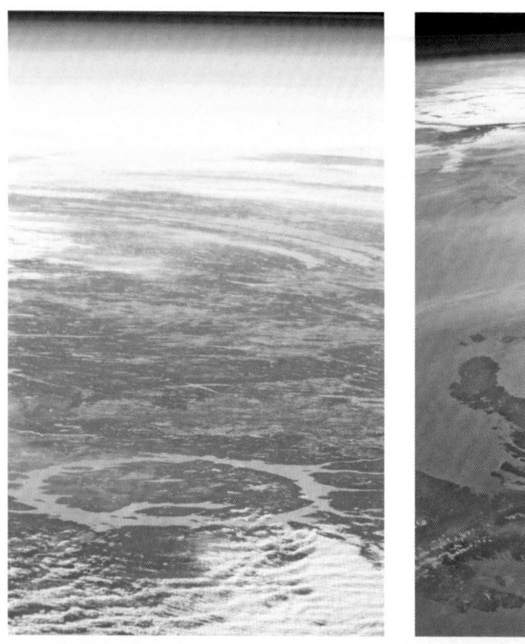

The view from the International Space Station is phenomenal—a visual onslaught of ever-changing light, texture and discovery. The Cupola's windows effortlessly offer up the familiar shapes of Earth in ever-unfamiliar and breathtaking ways. To simply look through the camera lens and press the shutter button is to see our world with both newfound understanding and respect. *(Credit: NASA/Chris Hadfield)*

heading into quarantine, I tried to figure out ways to acknowledge the costs to the people around me, ways to compensate them and ways to include them in any success I had.

For my second Shuttle flight I was in quarantine for my son Evan's 16th birthday. That's a big day in a teenager's life, a turning point when he could get a driver's license and was officially on his way to being considered an adult. But the hoopla surrounding the launch was overshadowing his birthday, and Evan was resoundingly unhappy about that. In quarantine, I was nicely isolated from his black cloud, so Helene was bearing the brunt. She did have visiting privileges, though, and did not hesitate to let me know.

Frankly, I just hadn't thought through the consequences of the timing in detail. At this late date, my only option was to try to make his birthday special in any way I could despite being holed up in quarantine. So I announced in some of the many phone interviews I did that we would be lighting the world's biggest candles—the Shuttle's rocket engines—to celebrate Evan's birthday. That made the news, so he heard it, as did everyone else who knew him. And just before we crawled into *Endeavour*, I held up a handwritten sign that said, "Happy Birthday, Evan!" Thankfully, the media noticed the gesture and ran with it as a nice family story. Evan was happy, or at least happier.

I learned my lesson. Before my last mission, I sat down with the calendar and planned: *Okay, I won't be around for Valentine's Day, so I'd better organize a card and get a gift right now, when I can plan and execute properly, so everything is in place on the actual day.* Forward planning was an easy way to show the people who made it possible for me to do my job that I didn't take them for granted. Making a flowery toast afterward, thanking your nearest and dearest for all their support, just won't cut it if, again and again, you've passed up opportunities to show appreciation in real time.

Early on, and aided considerably by the fact that suffering in silence is not considered a virtue by any member of my immediate family, I recognized that the only fair way to deal with the imbalances my job created was to anticipate crunch times at work and try to make it up to my family well in advance. Every year, when the kids were young, for instance, I took them on vacation by myself for 10 days—to Europe, the Grand Canyon, scuba diving in the Florida Keys—so that we could bond and Helene could have a break. Usually she just stayed at home and went to work, but she still says these "one-dish-in-the-sink" breaks were among the best of her life. And when I had a PR tour in an exotic destination, where I'd be giving speeches about the space program and explaining to media what we do and why it matters, we'd plan carefully so that at least one of the kids could come along and sightsee while I worked; in the evenings we'd have dinner together. Most PR tours are grueling: it's one interview and talk after another—six or seven events in a single day isn't unusual—then working on the plane on the way home. We did a few of these together, too, and that wound up being good for our family because afterward everyone really understood that when I traveled, I wasn't having a grand old time while they were stuck at home. A PR blitz is only fun if you make it fun, and in my experience that's hard to do without Helene and/or our kids there. Unfortunately, after a while, they caught on and demanded to see the itinerary and schedule before they'd agree to come.

My point, though, is that saying thank you every once in a while just isn't enough when you're demanding that other people make real sacrifices so you can pursue your goals. It's not only the fun, showy things like vacations that get the message across. You also have to be willing to do what you can to create the conditions that allow your partner the freedom to focus single-mindedly at times. It's not easy but it is possible with careful planning, regardless of the scope of your ambition or the demands of your job. Some astronauts wind up marrying other astronauts, after all, and starting families—and somehow, between stints in space, they're able to find a way to make it work.

When you have great backup, as I have always had, you can start to take it for granted or become selfish and just expect that your needs will take precedence. I've tried to guard against that by making sure that when I have any wiggle room in my schedule, Helene is the one who sets the agenda, whether it includes me or not. I also make a point of actively looking for opportunities to spend time together. On Sunday mornings, for instance, no matter what else is going on, Helene and I try to walk the dogs, then go get coffee and do the *New York Times* crossword puzzle together. Prioritizing family time—making it mandatory, in the same way that a meeting at work is mandatory—helps show the people who are most important to me that they are, in fact, important to me.

And it's not exactly unpleasant for me, either.

* * *

Time-honored astronaut traditions make us feel we're part of the tribe, and there were plenty of them during our final hours in quarantine. Some were less picturesque than others. The night before we launched, we gave ourselves an enema, followed, after a suitable interval, by another one. While this did not feel like my finest hour in space exploration, it was definitely preferable to soiling my diaper the next day. Afterward, a doctor took swabs of all parts of my body—behind my ears, my tongue, my crotch—to see if I had any infections, then rubbed me down with alcohol just in case I did.

On December 19, I put on my blue flight suit and headed to my final breakfast on Earth in 2012. This was more of a ritual than a meal. Tom, Roman and I restricted ourselves to clear fluids and a bit of gruel, eaten with sardonic awareness that we might see it all again in a few hours—post-launch nausea is common—and that we wouldn't have access to a private toilet until we got to the ISS two days hence. A bit later on, we went to a small room for a private toast with our spouses and a senior representative from each of the space agencies involved in our flight: the CSA, NASA and Roscosmos. We all said a few words, which those of us who would be flying a Soyuz toasted with ginger ale, not champagne, then everyone in the room sat down for a minute of silence. It's what Russians do before any voyage, whether they're going to space or to a friend's dacha, just a way of honoring the significance of the moment.

We were almost ready to leave the building we'd lived in for nearly two weeks. By way of farewell we signed the door of quarantine, adding our names to so many others, then walked down the hall toward the exit. Waiting there was a Russian Orthodox priest, dressed head to toe in black, and a helper, armed with a bucket of water. We stood in front of the priest, the backup crew right behind us, and he dipped something that looked a great deal like a horse's tail into the bucket, then flung water on us. Doused us, really, while he was blessing us.

Then we opened the door to walk out to the bus that would take us toward our spacesuits, our rocket, our next chapter. All our launch guests were lined up, waving flags and cheering, calling out goodbyes, stamping their feet. It was a bright, sunny day but bitterly cold, -25° or so. Lingering outside with wet heads seemed like an inherently bad idea, so after standing and waving for a minute outside the bus, we climbed inside and resumed waving. Through the window, I searched out my children, my wife, memorizing them, hoping they could see gratitude and love in my eyes, while the bus, heated to the point of stuffiness, slowly started rolling toward the compound's exit gate.

We were on our way.

8
HOW TO GET BLASTED (AND FEEL GOOD THE NEXT DAY)

Baikonur winters are never mild, but 2011 was particularly brutal. During the ceremonial events before the December launch that year, snow was blowing all over the place and an icy wind easily pierced the cloth, rubber and metal layers of the crew's spacesuits. By the time they reached their Soyuz, they were numb with cold. So for our own launch in December 2012, the Russians decided to take preemptive measures. They crafted pillowy white snowsuits for us, complicated multi-piece affairs that snapped on over our other gear like armor. Tom, Roman and I were a little dubious when we saw them. The diapers were bad enough. Now we had to wear giant duvets?

We were in the suit-up facility, a nondescript, industrial-looking building en route to the pad, and suit technicians had already helped us get into our Sokhols. In Russian, *sokhol* means "falcon," but these particular falcons can only fly inside a spaceship; like our bright orange Shuttle spacesuits, they are worn only to protect us during launch and landing, not for spacewalks. After pressure checks confirmed that our spacesuits had no leaks and could therefore keep us alive if the Soyuz depressurized in space, the suit techs began bundling us into our snowsuits. If nothing else, they provided comic relief. When we finally waddled out a side door of the building, we looked like Michelin men, overstuffed and big of rear end. To complete the picture, all three of us were clutching what looked like large aluminum lunch boxes, containing our ventilators.

It felt a bit as though we were still pretending to be astronauts en route to space, just as we had for years. But there was the bus, waiting to take us to the launch pad. And there were our families, friends and various officials from the Canadian, American and Russian space programs, waiting behind a rope for a glimpse of fully suited, honest-to-goodness, going-to-space-any-minute-now astronauts. The sky was cloudless and the sun shone brightly, but the air had a sharp bite. I heard my name being called and turned, catching momentary flashes of familiar faces in the crowd, then we were on the bus, waving goodbye. This time, it really was goodbye. We would not be seeing these people again anytime soon. Or perhaps ever. It was an inescapable fact that we were about to do something that's a whole lot riskier than getting on a plane. I was pretty sure I'd survive the day but still I didn't want to leave anyone with a last image that was either too somber or too flippant. Waving as the bus slowly pulled away, I hoped I looked exactly the way I felt: happy to be on my way, confident that I was ready to do my part, fully prepared for any outcome.

I knew for sure that I looked warm. After we'd been driving for 15 minutes or so, the windows had fogged up and the bus verged on unbearably hot. When the driver pulled over to the side of the deserted road, Roman, Tom and I were delighted to get out and breathe some fresh air. We also had a mission: to pee on the rear right tire of the bus, as Yuri Gagarin apparently had. Much is made of this as a tradition, but really, if you're going to be locked in a rocket ship, unable to leave your seat for quite a few hours, it's just common sense. However, we had a problem that previous crews had not: we had to figure out how to get out of our suits of downy armor. In the end the suit techs on board had to help us undo all the tricky fasteners they'd painstakingly closed not an hour before, so we were able to urinate manfully on the tire without spoiling our plumage. Female astronauts who bring little bottles of their pee to splash on the tire may feel just as self-conscious, but I doubt it.

Afterward, our backup crew came over from their bus—even this late in the day they traveled in a separate vehicle—to say goodbye. Hugs all around. They were happy to see us go: once we were off Earth, they'd move one big step closer to being prime crew. It would be their turn in six months.

Back on the bus, only a few minutes from the pad, our suit techs got busy cheerfully and efficiently lacing, buttoning and zipping us back into our snowsuits and checking our Sokhols; by undoing them to relieve ourselves on the bus tire, we had invalidated all the previous pressure checks. We were once again good to go by the time we pulled up to the pad for the farewell ceremony with the highest-ranking people in the Russian space industry. There were probably 50 technicians and officials waiting for us, including the head of Roscosmos and the head of Energia, the corporation responsible for building Russian spacecraft.

Roman got off the bus first. Naturally, as this was his country and he was commander of our Soyuz, he was the center of attention, which suited Tom and me perfectly. One of our goals was for Roman to emerge from our mission as a shoo-in to be commander of the ISS the next time he flew, so we took the attitude, "Don't mind us, we're with Roman." We followed him across the tarmac to our marks, where we stood and formally saluted the head of Roscosmos, Vladimir Aleksandrovich Popovkin. Next, the six highest-ranked dignitaries had the honor of escorting us to the steep stairs leading up to the rocket ship, each gripping one astronaut arm. The pair who'd won the coin toss proudly hung onto Roman, while the others, a little less proudly, helped me and Tom. Of course we didn't really need any assistance, but it was a nice symbolic gesture of support and, like the rest of the Russian rituals, imparted a sense of occasion to the proceedings.

This was the first time we'd seen our Soyuz vertical and ready for liftoff; the Russians consider it bad luck for a crew to see its own rocket ship on the pad any earlier than launch day. So two days before, only our families, friends and the backup crew had gathered before sunrise for rollout, which is a ceremonial unveiling of sorts: the Soyuz is transported from the vehicle assembly building to the pad by a humble low-tech train that labors down the tracks, seemingly in slow motion, while hypothermic onlookers cheer its placid progress. Kristin later told me that our guests' enthusiasm for the pre-dawn grandeur of rollout was only matched by their enthusiasm to get back on their heated bus. Once the sun had risen and temperatures had climbed to merely mind-numbing, they watched the rocket being lifted up from its horizontal position on the train and efficiently positioned on the pad by what looked like huge construction clamps. It was a good opportunity for them to see the Soyuz up close; on launch day, they watch from a viewing stand about a mile away—a safe distance, should anything go wrong, but close enough that the ground beneath their feet still trembles during liftoff.

As I was being escorted to the stairs I noticed that our rocket was encased in a thick sheath of ice, like an old-fashioned freezer in need of defrosting. Nothing to be concerned about, fortunately. Some version of the Soyuz has been flying for more than 45 years. As rocket ships go, it's one of the most reliable and durable in the world, and can safely launch in just about any weather.

I was the first one up the stairs and as I climbed, the chief of Energia gave my rear end a swift, friendly kick—the Russian equivalent of "break a leg." It's the symbolic push-off to launch and not at all unpleasant when you're as well padded as I was. Partway up, I stopped and turned, as did Tom and Roman, to wave one last time. It was a Kodak moment—three guys, off on an amazing adventure!—and one we decided, by unspoken agreement, to keep mercifully brief. We had somewhere to go.

* * *

Fifty percent of the risk of a catastrophic failure during a long-duration space mission occurs in the first 10 minutes after liftoff. Per second, it's the most dangerous phase of space flight. So many complex systems are interacting that changing a single variable can have a huge ripple effect, which is why we train so long and hard for launch: you have to know how the dominoes might fall, and be ready to do the right thing, in all different kinds of scenarios. Often you have only a few seconds to react. You feel the pressure even during training. No one wants to die in the sim—it doesn't look good.

Sometimes the clues that something is going wrong are very subtle. On the Shuttle, for instance, four computers, all running the same software simultaneously, controlled the vehicle. Regular laptops on Earth occasionally freeze or have software glitches, but the odds of computer problems significantly increase in space, thanks to the stresses of launch: vibration, acceleration, changing electrical current and fluctuating heat. That's one reason these four computers were linked, so they could constantly compare what they were doing. If one did something dumb, the other three could overrule it and shut it down. But if even a tiny timing error developed, two of the four could split off and go rogue—giving the vehicle directions that contradicted what the other two computers were instructing it to do—with no one to break the tie and vote on which pair was right. The main way to figure out if we had a "two-on-two set split" was to monitor the pattern of some lights on an overhead panel, while we were trying to do a million other things, too. But it wasn't a task we could afford to overlook. If the Shuttle responded to conflicting directions by turning suddenly during launch, say, the vehicle could simply break up midflight, unable to withstand the structural stress caused by rapid changes in aerodynamic flow. To avert catastrophe, we'd have had to recognize a bad set split instantly and respond within seconds. Both the pilot and the commander would, simultaneously, have had to override the four main computers and activate the backup computer, which was relatively primitive but could, in an emergency, get the Shuttle back to Earth.

During a Shuttle launch, we also needed to recalculate, constantly, how and when to shut the engines down manually in case of an emergency. You couldn't just turn them off abruptly while accelerating; picture sailing down the highway at 80 miles an hour, then suddenly shutting off your motor—it wouldn't be a good idea for the car. Or for you. Well, the risk is exponentially greater when you're traveling 8,000 miles an hour and huge turbo pumps, powerful enough to drain a swimming pool in less than 30 seconds, are pushing fuel into the motor. If a Shuttle engine wasn't shut down gracefully and gradually, it would blow up. So during launch, we spent a lot of time working a hypothetical problem: how, if something went wrong, we'd throttle back. In fact, on two separate Shuttle missions, the crews did have to shut down an engine. But because they'd been trained so well to think through interconnected webs of problems very quickly and calmly, those shutdowns were nonevents and both flights continued as planned. That's why, in all likelihood, you've never heard of them until now.

The Soyuz is a much simpler vehicle to operate and it is automated: if something goes terribly wrong, the chances of survival are much better than they were on the Shuttle because the re-entry capsule where the crew sits during launch automatically separates and is thrown clear. This is what happened in 1983, two seconds before a Soyuz exploded on the pad during the final countdown; the crew survived. In 1975, after a serious booster malfunction partway through ascent, pyrotechnics automatically fired to blast the crew's capsule free of the rocket; as it fell back to Earth, its parachutes deployed properly, right on schedule. However, that Soyuz crash-landed in a hilly, remote area and promptly began to roll down a snowy slope, coming to a stop at the edge of a steep cliff only because the parachute snagged on some vegetation. The crew lived to tell the tale. Only once has a parachute failed: on the very first Soyuz flight, in 1967. Vladimir Komarov, the cosmonaut on board—he was the only one; it was a test flight—was killed, the first inflight fatality in the history of human space exploration. Since then, thankfully, both the vehicle and its parachutes have been eminently reliable.

Our crew felt confident that even in the case of an engine failure, we'd almost certainly survive. However, not all engine failures are equal, not even on a highly automated rocket ship. On the Soyuz, one of the worst times for this to happen would be just after the first two minutes in flight, when the vehicle is way up high but not yet going all that fast. You'd fall straight back down. If the Soyuz comes back to Earth horizontally, it bumps along the atmosphere, like a stone skipping across the surface of a pond, slowing down before coming to a stop. But if it's plummeting vertically, it's like a stone being dropped into a pond from a great height. The rocket ship would hit the thick air of the atmosphere all at once, creating deceleration forces up to 24 g—survivable, but extremely punishing for both humans and spacecraft. The Soyuz commander would have about four seconds to make a crucial difference: by pushing buttons on the manual control handle, it's possible to override some of the automatics and roll the re-entry capsule to an orientation that reduces the g-load by as much as 8 or 10 g. While 14 or 16 g is still a wicked load, it is a whole lot better than 24. So Roman practiced doing this in the sim and we all talked about it every time, just in case.

Really, we had practiced doing everything so thoroughly—and had thought so much about what could kill (or just maim) us next—that we felt, heading into launch, prepared for just about anything. We had had countless opportunities to zero in on our weaknesses and try to improve in those areas, as well as countless opportunities to develop and practice new skills. The mental and emotional toughness necessary to handle the pressure and stress of launch had developed during that slow, arduous process. Our core skill, the one that made us astronauts—the ability to parse and solve complex problems rapidly, with incomplete information, in a hostile environment—was not something any of us had been born with. But by this point, we all had it. We'd developed it on the job.

Being well prepared didn't mean we were jaded, though. For me, as for anyone who's embarking on any kind of hard-earned mission, launch felt both daunting and wildly exhilarating. My first time, I'd felt pent-up excitement mixed with a rookie's earnest desire to prove myself. My second launch had been different; then, I'd been gripped by an intense sense of purpose, knowing that the correct installation of Canadarm2 was crucial for the future of the ISS. Before this third launch, the last of my career, I felt I was dancing with the devil I knew, confident in myself, my crew and our spaceship. It was a strange combination of feeling peaceful and rueful, almost, about what it had taken to get to this point. I was determined to make the most of every moment of this incredible journey, to engrave its details on my memory. I would have to. I'd never get another chance.

* * *

The Soyuz is so small that it makes the Shuttle seem almost cavernous. A Dodge Caravan has about 163 cubic feet of space; the Soyuz has 265 cubic feet of living space—theoretically. In reality, a lot of that space is taken up by cargo and gear that's been lashed down and secured for launch. In any event, it's not a lot of space for three full-grown adults to share for a few days. But during launch, we have even less elbow room because we are confined to the re-entry module, which is also the only part of the Soyuz that survives the return to Earth. On our way home we jettison the other two: the service module, which houses the instruments and engines, and the orbital module, which provides additional living space once we are on orbit.

When Tom, Roman and I reached the top of the stairs, a technician hustled us into a smallish elevator that whirred and clunked as it ascended, then deposited us into a cramped booth with a hole in the side, reminiscent of an igloo. We took off our white padding and then, one at a time, crawled on our hands and knees through the hole and into the orbital module. I was the left-seater, the pilot, so I went first because my seat was the least accessible. After launch, the orbital module becomes our living room, essentially, but it was startling to see that it had already been filled almost to the ceiling with a hodgepodge of equipment and supplies. It looked like a station wagon jammed to the roof for a long cross-country trip. I noticed my checklists perched on top of a 3-foot-high tower of stuff, but I was already focused on lowering myself carefully into the re-entry module, where we sit for launch and landing. I didn't want the big regulator valve on the front of my Sokhol to scratch up the hatch.

Once I was in my seat, which had been custom-molded to my body in order to absorb the shock of landing, our strap-in technician, Sasha, climbed in to help me get belted in tightly. You might think that in such a tiny vehicle the tech would have to be small and wiry, but Sasha was a beefy guy with the build of a nightclub bouncer. After he got me wedged in securely, I thought to ask him to hand me my checklists. He said he would, then went back up to the waiting room without giving them to me.

My job was to start checking the systems, to make sure everything was working, but . . . I needed those checklists. I called up but no one responded because they were busy helping Tom. Great. I'd have to start up the Soyuz from memory. No. Bad idea. After Tom got settled into his seat and Sasha came down to strap him in, I reminded him that I needed my data file. Sasha said, "Oh, the guy up there says you don't need it yet." *What* guy? And the file didn't belong to "the guy," whoever he was. It was mine. But I couldn't move. By the time Roman got in, the re-entry capsule was so jammed that Sasha couldn't assist him, so Tom and I did, and then Roman looked around and wondered where his checklists were. Finally, at that point, they passed them down to us. They wanted to wait, I guess, until someone really trustworthy—a commander, not just a garden-variety astronaut—was in place.

As it turned out, I needn't have worried. We had plenty of time to work through our lists and verify that everything was functioning as it should be. From all our sims, it felt very familiar: same sort of seats, same sort of tasks, same sort of checklists. Even the same voice was coming through our headphones: that of Yuri Vasilyevich Cherkashin, our instructor. Everything looked and felt just about as it always had, all the times we'd practiced—right up until Roman threw the big lever and the little lock, closing the re-entry hatch from our side, and Sasha closed it from his, saying, "*Schastlivovo puti.*" Bon voyage.

Or, to put it another way: time to wait some more. There was still a lot to do before we could lift off, pressure checks being the most crucial. We had to be sure the seals on our vehicle were tight. They were. Then we had to check that our Sokhols were still hermetically sealed and, in the event of a leak in the Soyuz, could essentially become our own individual spacecraft and buy us time to get back to Earth. Without them, we'd die quickly but not painlessly, starved of oxygen. First we closed and locked our helmets, reminding each other that we needed to hear *dva zaschelkami*—two clicks—then we clamped down on our regulators until our Sokhols inflated like balloons. It's not the best feeling in the world—it's hard to keep your ears clear—but within about 25 seconds we knew we could trust our spacesuits in an emergency. We waited the full, prescribed three minutes for the ground to be satisfied as well, then popped our helmets open and I turned off the oxygen supply. We already had plenty in the capsule—no need to increase the risk of fire.

I carefully tried all the displays—there are about 50, covering everything from speed/altitude to the ship's oxygen system to the mathematical summaries of orbital targets—to be sure they worked the way they had in the sim. They did. We had controlled everything we could control. Our vehicle was healthy. We'd done everything on our checklists. Our suits worked. I'd been sitting in the Soyuz about two hours now, knees crunched up almost to my chest. The backs of my knees did ache a bit from the last pressure check, and my lower ribs were reminding me where I'd broken them years ago, water-skiing at Pax River. But other than that, I felt good. Normal. Hungry in fact, as did Roman and Tom. It was almost dinnertime, after all, and we'd eaten almost nothing all day. We'd have to wait a few more hours.

Outside they were moving the gantry—the portable structure with the stairs, elevator and ingress room—away from our rocket ship. Forty minutes or so to go. Yuri had asked us to choose some songs we'd like to hear while we were waiting, and he'd also chosen a few for us. He knows us well. As the music began to play, we were smiling, explaining the particular significance of each song to one another. For Tom, there was classical guitar—he's a good guitarist and was planning to practice on Station. For me, my brother Dave's song "Big Smoke," linking family, history, music and my own current location, atop what would soon enough become a major smokestack. For Roman, the youngest of us, some rock music, the bouncy kind that makes you want to dance in your seat, even when you're strapped in so tightly that it's difficult to move. I'd asked for "If You Could Read My Mind," my favorite Gordon Lightfoot song; thoughtful and soaring, it always brings me peace. And since, according to the Mayan calendar, we were just two days away from the end of the world, I'd also picked Great Big Sea's high-speed rendition of "It's the End of the World as We Know It (and I Feel Fine)." We heard U2's "Beautiful Day" and Depeche Mode's "World in My Eyes," which starts, "Let me take you on a trip / Around the world and back / And you won't have to move / You just sit still."

Sit still and remain calm is exactly what we were trying to do as the minutes ticked by and the sun sank lower in the sky. We were scheduled to lift off right after sunset. We didn't want our hearts to start racing with excitement five minutes before launch. Underneath the Sokhol we wear something like a training bra that has electrodes to transmit medical data to the ground. None of us wanted to give the team of doctors who were monitoring our every heartbeat anything at all to worry about. Especially not me, not now, not after what I'd gone through to be cleared to fly. On my ascent checklist I'd actually penciled in a reminder: "Be calm. Medical parameters."

Sweat the small stuff. Without letting anyone see you sweat.

A few minutes before launch, with the Beatles' "Here Comes the Sun" playing, we turned to the launch page: there was just one, covering everything from ignition to cut-off. Incredible, really, for such a complex series of events, but we needed to be watching our displays, hawking it. Anyway, it was a given that we knew the boldface cold. "*Miakoi posadki*," Yuri said, by way of farewell. "Soft landings." It's what we wished for, too.

The smaller, outer engines started lighting about 30 seconds before liftoff, so that Ground Control could really be sure everything was up and running properly before lighting the engines that had enough power to push us off the Earth. It was a way of hedging their bets and, for me and Tom, a gentle initiation to the Soyuz. We felt a rumbling sensation but, unlike the Shuttle, no twangs, no swaying. The Shuttle engines stuck out to one side, so when they lit, their force pulled and bent the spaceship. The Soyuz engines, however, are symmetrical, firing up through the vehicle's center of gravity, so while there's a steadily mounting vibration, there's no off-center motion, no sudden, explosive announcement that you are leaving the planet.

The rumble of power just got stronger and more insistent as we heard the countdown in Russian through our headsets and then, "*Pusk*." Liftoff. It was a very different sensation than my two Shuttle launches, much more gradual and linear as the vehicle burned off enough fuel to lighten for liftoff. The initial acceleration didn't feel all that different from just sitting on the ground. We knew we were leaving the pad more because of the clock than the sensation of speed.

From the viewpoint of those watching in the stands, those first 10 seconds of the launch were agonizingly slow. Kristin later admitted she'd been terrified, so much so that she hadn't wanted to take any photos or take her eyes off the Soyuz for a second. Compared to a Shuttle launch, the rocket ship seemed to hover above the pad just a little too long. One guest likened it to the ultimate bench press, saying it looked as though an unseen weight lifter was standing underneath, straining mightily to push the vehicle off the ground, but failure was always an option.

Inside the vehicle, however, we were full of anticipation, not dread: ready for this machine to do its work. It was like being a passenger in a big locomotive, but one who can throw the emergency hand brake if necessary. We had some degree of control. The challenge was knowing if and when to assert it. Within a minute, we were pushed down in our seats more and more heavily. Initial ascent felt purposeful but smooth, a little like being on a broomstick that an invisible hand was calmly steering a bit to the left, then a bit to the right, back and forth. The rocket ship was self-correcting its attitude as we ascended and the wind and jet stream changed.

The ride got less smooth as it went on, though. As our first-stage engines cut off and the boosters exploded off the side, there was a noticeable change in vibration and a decrease in acceleration—not speed, that was always increasing. We were thrown forward and then steadily pushed back again as the Soyuz, lightened, roared upward. This tail-off, lurch-forward motion was repeated when the second-stage engines separated, and as the third-stage engines lit, the ones that would take us to orbital speed, we were slammed back even more definitively. But that was a very good thing to feel, because a year before, the third stage hadn't lit on an unmanned Progress resupply vehicle and it had crashed in a sparsely populated region of the Himalayas. If that happened to us and the Soyuz parachutes deployed, it would be days before anyone found us. We'd all done winter survival training in remote areas to be prepared for just such a scenario, so we had a good idea just how miserable those days would be. At this point in the year, we'd no doubt wish we still had our Michelin Man suits.

The whole way up, we breathed a little easier as each important milestone passed. But it was not a nerve-wracking process. Approaching certain thresholds we knew it was possible that something really bad might happen, but we also had a plan for what each of us would do. We were wide awake and ready to take action. If anything went drastically wrong, like the engines didn't cut off on time, I would throw a switch and press two emergency buttons to fire the explosive bolts that would blast our capsule away from the rocket. I would have five seconds to assess what had gone wrong and take the appropriate actions. The three of us had gone over who was going to do what, with whose permission, again and again. We had agreement that if X didn't happen within Y seconds, I was going to activate contact separation. The left-seater is the only person who can even reach those buttons. I had raised the lids that normally cover them so I was ready to press at any moment, and it was a wonderful moment when I could close those lids.

Nine minutes had passed. Our third-stage engines had cut off, the Soyuz had separated, and its antennae and solar panels had deployed. Flight control was about to switch from Baikonur to the Russian Mission Control Center in Korolev, a suburb of Moscow.

Every crew brings its own small, tethered "g meter," a toy or figurine we hang in front of us so we know when we are weightless. Ours was Klyopa, a small knitted doll based on a character in a Russian children's television program, courtesy of Anastasia, Roman's 9-year-old daughter. When the string that was holding her suddenly slackened and she began to drift upward, I had a feeling I'd never felt before in space: I'd come home.

* * *

The life of an astronaut is one of simulating, practicing and anticipating, trying to build the necessary skills and create the correct mind-set. But ultimately, it's all pretend. It's only when the engines shut off and you check that you're pointed the right way and going fast enough that you can acknowledge, "Hey, we made it. We're in space." Maybe it's not unlike childbirth in that the end result has been in your head all along; you've read the books and seen the pictures, you've prepared the baby's room and taken the Lamaze classes, you've got a plan and think you know what you're doing—and then, suddenly, you're confronted with a squalling infant, and it's wildly different.

In 1995, I was the only rookie on our crew. I didn't want to show up in space with that lost, first-day-on-the-job feeling of, "Well, now what am I supposed to do?" We were only going to be up there for eight days, total. I didn't want to spend any of them feeling—and being—useless. So while I was still on Earth, I thought through, sequentially and in detail, exactly what would happen as soon as we reached orbital speed, and I came up with a list of things I ought to do. I'm not talking about big, vague goals like "demonstrate leadership abilities." I'm talking about real nuts-and-bolts stuff, like putting my gloves and checklists into the mesh helmet bag, then collecting the foam head supports from everyone's launch seats and putting them in the "Bones Bag" for items unneeded on the return flight.

Having a plan of action, even really mundane action, was a huge benefit in terms of adaptation to a radically new environment. I'd never experienced zero gravity before, for instance. I "knew" exactly what it would feel like, from all my training and studying—only, I didn't really know at all. I was accustomed to being pulled down to the floor by gravity, but now felt I was being pulled up to the ceiling. It was one thing to sit in my seat and watch stuff float around, but quite another to get up and try to move around myself. It was a profoundly disorienting form of culture shock, literally dizzying. If I moved my head too quickly, my stomach flipped, sickeningly. My to-do list gave me something to focus on aside from my own disorientation. When I did the first thing on my list and it worked, and then the second and third things worked, it really helped me find my footing. I developed some momentum; I didn't feel so lost.

It's obvious that you have to plan for a major life event like a launch. You can't just wing it. What's less obvious, perhaps, is that it makes sense to come up with an equally detailed plan for how to adapt afterward. Physical and psychological adaptation to a new environment, whether on Earth or in space, isn't instantaneous. There's always a bit of a lag between arriving and feeling comfortable. Having a plan that breaks down what you're going to do into small, concrete steps is the best way I know to bridge that gap.

On the Soyuz, we didn't have to rack our brains to come up with a list. There were a lot of practical housekeeping matters to take care of as soon as we were on orbit, and the confined space forced us to choreograph them carefully. First and most important: checking the pressure. Once we were certain the automatic systems were working and the maneuvering thrusters' fuel lines were full, we shut off the oxygen supply and measured the pressure in the re-entry and orbital capsules for an hour. If it fell even a little we'd have to turn around and head for one of the backup landing sites around the world—or, depending on the severity of the situation, head for anywhere at all and hope we wouldn't crash down in someone's backyard.

But our ship was airtight, so Roman opened the hatch to the orbital module and floated up to get out of his Sokhol. We had to take turns: there just isn't enough room in the Soyuz for three adults to climb out of their spacesuits at the same time. Getting out is easier than getting in, but it's still awkward, not least because the inside of the Sokhol is, by this point in the journey, distinctly clammy and feels the way a rubber glove does after you've been wearing it for a while. You actually have to attach the suit to a ventilator for a few hours until it dries out.

The next thing to come off: the diaper. Pride compels me to report that I've never used mine, but those who have are particularly happy to remove it. Now we were down to just our long underwear—100% cotton because, in the event of a fire, it chars nicely without melting or burning. Most astronauts stay in their long johns until it's time to dock with the ISS, reluctantly changing only because we know there will be TV cameras and looks of horror on the other crew's faces if we greet them decked out in dirty underwear. The approach to hygiene on the Soyuz is about what you'd expect on a camping trip. Decorum is a relative concept on a vehicle that size; there's no bathroom, for instance, so if you need to go, your crewmates simply look away politely while you pick up a thing that looks a bit like a DustBuster with a little yellow funnel attached. It's simple to use: turn the knob to "on," check that the airflow is actually working, then hold it up close so you don't get pee everywhere. A quick wipe with a piece of gauze and the funnel is dry.

As soon as I got out of my Sokhol I took anti-nausea meds. Feeling nauseated is inevitable during the first day or so in space because weightlessness completely confuses your body. Your inner ear no longer has a reliable way of judging up from down, which throws your balance out of whack and makes you feel sick. In the past, some astronauts vomited throughout their entire flights; their bodies just never accepted the absence of gravity. I knew mine would eventually adapt, but I didn't see the point of being sick my first few days in space, so I took the medication that was on offer and didn't eat very much.

I also didn't spend a lot of time gazing out the window at first. Unlike the Shuttle, which was powered by fuel cells, the Soyuz is solar-powered; to keep its solar arrays pointed at the sun, the vehicle spins like a chicken on a rotisserie barbecue. Outside the window, then, what you see is Earth, tumbling over and over, which is hard to look at when your stomach is unsettled. I waited until we were going to do an orbit adjust burn, in which case we'd maintain a stable attitude, before admiring the view.

That first evening we did two orbit adjust burns, firing the engines to climb higher toward the ISS. It's one of the most critical phases of flight on a Soyuz, because an error could rapidly put the rocket ship into an orbit where it would never reach Station at all. "There's nothing more important than what you're doing right now" is a standard astronaut adage that's never more true than when an engine is firing. All three of us stopped and stared, unblinking, at the display readings for fuel pressure, steering and propellant flow—anything that would tell us whether an engine was misbehaving. Collectively, we shared a hair-trigger reflex, but it was my job to act on it and push the appropriate immediate action buttons—there are 24, covered with small flip-lids to prevent inadvertent pushes—to shut down an errant engine manually and switch to backup thrusters, if necessary. But it wasn't. Behind us, a trail of burning snowflakes from the firing sparkled away into the night.

We'd checked all our thrusters and tested the computers, hand controllers and rendezvous radar that we'd need for docking with the ISS. Only a few hours into our journey, we'd done just about everything we had to do. Floating past the Soyuz TV screen, I noticed we were over the Pacific, off the Chilean coast. At the window, I saw a few lights: fishing boats, I thought. Then they resolved themselves: the Southern Cross. I was looking at a constellation in the night sky, not the sea! It was a strange delight to be that disconcerted while simultaneously at ease.

I realized I was tired. Very tired. I unrolled my sleeping bag, pale green with a white liner, and tied the four corners loosely to the metal rings on the sides of the Soyuz with the strings I found in the bag pocket. I didn't want to drift around going bump in the night. It was chilly in the capsule now. Fully clothed and wearing calf-length down slipper boots, I got into my bag, stuck my arms through the side holes, pulled on the built-in hood and zipped up. Floating inside, slightly curled like a baby in the womb, I fell asleep almost immediately, with Tom beside me and Roman a few feet away in the re-entry capsule. It was my first night in space since April 2001. Expedition 34/35 had begun.

* * *

Getting up to the ISS really doesn't take that long: you could make it there from Earth in less than three hours if you had to, and recently, several crews have done so, in the interests of efficiency. But we were allotted more than two days, as Soyuz crews usually have been, and I was glad of that time to ramp down from the adrenaline of launch and get used to the reality of being in space. On Station, we'd be conducting and monitoring scientific experiments, maintaining and repairing the spaceship itself, communicating constantly with Mission Control—the schedule would be packed.

A full day in limbo, before all that started, gave us a chance to adapt and reflect, almost undisturbed. On the Soyuz, unless you're directly over Russia, you don't have communication with the ground. A few times a day, then, we'd give Mission Control in Korolev a summary of the status of the vehicle, and they'd give us any data we needed for rendezvous and docking. Otherwise: peace and quiet. We were alone.

I woke at 5:30 DMT (Decreed Moscow Time) and quickly calculated: seven hours of sleep. I felt rested, though puffy-faced and congested—typical adaptation symptoms. My joints ached somewhat after being motionless for so many hours during launch and I had a bit of a headache, but the main thing I was aware of was a quiet sense of joy.

The night before, digging through the storage locker by his seat in the re-entry capsule, Tom had discovered cards from our spouses. I'd saved mine, tucking it away in my left leg pocket. Now, while the sun was coming up, I wanted to read it. As I opened the envelope, two small paper hearts floated out, turning slowly and catching the sun's rays. I trapped them carefully in my hand and held them as I read Helene's words. I decided those hearts would keep me company in my small sleeping pod on the

ISS over the next five months, delicate and vivid reminders of my life on Earth.

By this point Tom was waking too, so we rooted around for nasal spray and anti-nausea pills in the large toolbox-sized metal box called, prosaically enough, container #1. Roman was also stirring. We took turns peeing, then retrieved breakfast: canned cheese bread, dried fruit and a juice box. Coffee would have been nice, we agreed, but we'd have it soon enough, in pouches, on the ISS.

Roman was already moving quickly and energetically, smoothly efficient, as if his last long stay in space had been only yesterday. This Soyuz was his, and he treated it with proprietary care and respect. He soon settled down to watch the old Soviet comedies from the 1960s that Energia had loaded on his iPod. Tom was unobtrusive, solicitous and clearly happy to be back in space. He moved more deliberately and patiently, ever helpful. I felt relaxed and lazy, like a bubble in a languid stream. I took off my Omega Speedmaster watch to play with it in weightlessness. With a little push it became a metal jellyfish, the strap pulsing in and out like a living thing.

My body was starting to remember zero gravity, which, when you get used to it, is like being on the best ride at the fair, only it never stops. You can flip and tumble and float things across the spaceship, and it never gets old. It's just a constant, entertaining change of rules. And as my vestibular system adapted during our day of downtime, I started to be able to look out the window for longer and longer periods of time. The world was rolling by underneath, every place I'd ever read about or dreamed of visiting streaming past. There was the Sahara, there was Lake Victoria and the Nile, snaking all the way up to the Mediterranean. Explorers gave their lives trying to find the source of the Nile, but I could detect it with a casual glance, no effort at all.

The night sky was beautiful, too: fine-spun necklaces of countless tiny lights dressed up the jet-black cloak covering Earth. Looking out on the second day of our mission, I became aware that in the far distance, there was a distinctive-looking star. It stood out because, while all the other stars stayed exactly the same size and shape, this one got bigger and bigger as we got closer to it. At some point it stopped being a point of light and started becoming something three-dimensional, morphing into a strange bug-like thing with all kinds of appendages. And then, isolated against this inky background, it started to look like a small town.

Which is in fact what it is: an outpost that humans have built, far from Earth. The International Space Station. It's every science fiction book come true, every little kid's dream realized: a large, capable, fully human creation orbiting up in the universe.

And it felt miraculous that soon we'd be docked there, and the next phase of our expedition would begin.

9 AIM TO BE A ZERO

A FRIEND OF MINE was once in a crowded elevator in Building Four South at JSC in Houston when a senior astronaut got on and just stood there, visibly impatient, waiting for someone to divine that he needed to go to the sixth floor, and push the button. "I didn't spend all those years in university to wind up pushing buttons in an elevator," he snapped. Incredibly enough, someone did it for him. This incident made such a big impression on my friend that I heard about it, and probably a lot of other people did, too. For me, it was a cautionary tale about the pitfalls of ever thinking of yourself as An Astronaut (or A Doctor, or A Whatever). To everyone else, you're just that arrogant guy on the elevator, craving significance.

Over the years, I've realized that in any new situation, whether it involves an elevator or a rocket ship, you will almost certainly be viewed in one of three ways. As a minus one: actively harmful, someone who creates problems. Or as a zero: your impact is neutral and doesn't tip the balance one way or the other. Or you'll be seen as a plus one: someone who actively adds value. Everyone wants to be a plus one, of course. But proclaiming your plus-oneness at the outset almost guarantees you'll be perceived as a minus one, regardless of the skills you bring to the table or how you actually perform. This might seem self-evident, but it can't be, because so many people do it.

During the final selection round for each new class of NASA astronauts, for example, there's always at least one individual who's hell-bent on advertising him- or herself as a plus one. In fact, all the applicants who make it to the final 100 and are invited to come to Houston for a week have impressive qualifications and really are plus ones—in their own fields. But invariably, someone decides to take it a little further and behave like An Astronaut, one who already knows just about everything there is to know—the meaning of every acronym, the purpose of every valve on a spacesuit—and who just might be willing, if asked nicely, to go to Mars tomorrow. Sometimes the motivation is over-eagerness rather than arrogance, but the effect is the same.

The truth is that many applicants don't have any realistic idea of what it means to be an astronaut. How could they? In the movies, astronauts are not toiling over Russian vocabulary work sheets. They're superheroes. Even the most level-headed among us have been influenced to some degree by that image. I know I was. So one purpose of that week at JSC is to dispel any comic-book notions about what working for NASA is really like. And some people do take a look around and run for the hills.

Those who aren't scared off are, in between familiarization sessions and tours, put through their paces. We give them an intelligence test and an aptitude test for manipulating robotic equipment such as Canadarm2, which requires the ability to visualize in 3-D (it's quite tricky). We even suspend applicants in simulated zero gravity to get a sense of their hand-eye coordination. Other assessments, like figuring out who plays well with others, are less formal. Applicants certainly know during the social mixer with astronauts from the office that we are evaluating them as potential crewmates, but they probably don't know who else has input. One Chief Astronaut used to make a point of phoning the front desk at the clinic where applicants are sent for medical testing, to find out which ones treated the staff well—and which ones stood out in a bad way. The nurses and clinic staff have seen a whole lot of astronauts over the years, and they know what the wrong stuff looks like. A person with a superiority complex might unwittingly, right there in the waiting room, quash his or her chances of ever going to space.

Which is a good thing, really, because anyone who views him- or herself as more important than the "little people" is not cut out for this job (and would probably hate doing it). No astronaut, no matter how brilliant or brave, is a solo act. Our expertise is the result of the training provided by thousands of experts around the world, and the support provided by thousands of technicians in five different space agencies. Our safety depends on many tens of thousands of people we'll never meet, like the welders in Russia who assemble the Soyuz, and the North American textile workers who fabricate our spacesuits. And our employment depends entirely on millions of other people believing in the importance of space exploration and being willing to underwrite it with their tax dollars. We work on behalf of everyone in our country, not just a select few, so we should behave the same way whether we're meeting with a head of state or a seventh-grade science class. Frankly, this makes good sense even if you're not an astronaut. You never really know who will have a say in where you wind up. It could be the CEO. But it might well be the receptionist.

If you enter a new environment intent on exploding out of the gate, you risk wreaking havoc instead. I learned this the hard way in graduate school, when we were in the lab designing low-pressure fuel pumps. We tracked our progress using different dyes, and at the end of the first day, we had an array of jars filled with leftover dye. I very efficiently took charge and poured them all down the drain in the corner of the room. Why bother asking questions? I already knew what needed to be done. Well, as it turned out, that particular drain was actually part of the lab's data collection system and therefore had to be kept spotless. The professor who ran the place couldn't believe I'd dumped dye all over it. Now the whole system had to be purged and purified, which meant a lot of extra work for him and other people. I'm sure that if he connected the dots today, he'd say, "*That* guy became an astronaut? But he's an idiot!"

When you have some skills but don't fully understand your environment, there is no way you can be a plus one. At best, you can be a zero. But a zero isn't a bad thing to be. You're competent enough not to create problems or make more work for everyone else. And you have to be competent, and prove to others that you are, before you can be extraordinary. There are no shortcuts, unfortunately.

Even later, when you do understand the environment and can make an outstanding contribution, there's considerable wisdom in practicing humility. If you really are a plus one, people will notice—and they're even more likely to give you credit for it if you're not trying to rub their noses in your greatness. On my second National Outdoor Leadership School survival course I shared a tent with Tom Marshburn, my crewmate on Expedition 34/35. Tom is the ultimate outdoorsman: a vastly experienced mountaineer, he's summited on several continents and also walked the Pacific Trail—alone—from Canada to Mexico, covering more than a marathon's distance each day. And yet during our course in Utah, he never imposed his expertise on anyone or told us what to do. Instead, he was just quietly competent and helpful. If I needed him, he was there in an instant, but he never elbowed me out of the way to demonstrate his superior skills or made me feel small for not knowing how to do something. Everyone on our team knew that Tom was a plus one. He didn't have to tell us.

* * *

So how do you get to be a plus one, someone who adds value? I wasn't certain when I was training for mission STS-74 in 1995, so as I mentioned earlier, I watched Jerry Ross, the most experienced astronaut on our crew, to see how he did things. After a while, I noticed that he was regularly coming into the office an hour early and quietly plowing through our commander's inbox, taking care of all the administrative details himself so the commander could focus on the important matters. I'm sure Jerry wasn't asked to do this, and he never mentioned it, let alone expected any recognition for it. He was voluntarily pushing the elevator buttons for someone else, so to speak, without fanfare or resentment. It was classic expeditionary behavior, putting the needs of the group first.

It was also a big part of what made him a plus one on our crew. Not only did he bring a wealth of experience and knowledge, but he conducted himself as though no task was beneath him. He acted as though he considered himself a zero: reasonably competent but no better than anyone else.

That made a lasting impression on me. Especially when I'm entering a new situation and don't yet have the lay of the land, I think about how to aim to be a zero and try to contribute in small ways without creating disruptions. Approaching the ISS in

December 2012, our crew talked about how to do this. Leaving Earth, we'd been treated like conquering heroes. But when we opened the hatch and floated into the ISS, we'd just be the new guys, the ones who didn't know where anything was. We'd be joining a crew of three people who'd been working and living on the ISS for months; they'd have developed their own shorthand for communicating, their own ways of doing things, their own routines. They'd probably be happy to see us—fresh supplies!—but also a little wary. What if we put trash in the wrong place or inadvertently ate the last pouch of peach ambrosia that someone had been saving for a treat?

We could also create bigger problems. When you first come into the Station after a few days of confinement in a Soyuz, you're disoriented and clumsy (not least because you're probably pretty anxious to get to a more-or-less private bathroom). It's like being a baby bird and not quite knowing how to fly yet. You might float past what just looks like a bunch of junk on the wall but is actually a biological experiment—bump it accidentally and years of science (and someone's life work) might be destroyed. This actually happened during my second mission: someone on our crew brushed up against an experiment as we entered the ISS, wiping out a whole month's worth of data.

The ideal entry is not to sail in and make your presence known immediately. It's to ingress without causing a ripple. The best way to contribute to a brand-new environment is not by trying to prove what a wonderful addition you are. It's by trying to have a neutral impact, to observe and learn from those who are already there, and to pitch in with the grunt work wherever possible.

One benefit of aiming to be a zero: it's an attainable goal. Plus, it's often a good way to get to plus one. If you're really observing and trying to learn rather than seeking to impress, you may actually get the chance to do something useful. For instance, before I'd ever flown in space, I was in a Shuttle entry sim with two very experienced astronauts. I was in student mode, keeping my eyes open and my mouth shut, when the commander reached up to turn something on. Because I was watching so closely, I knew without a doubt that he was about to press the wrong button. So I said, "Wait, that's not the right one." No big deal. He readjusted, the sim went on and I didn't say anything else about it nor did anyone else. A few months later, though, we happened to be at the Cape together for a launch, talking to the head of JSC, when with no prompting or warning, the commander began extolling my powers of observation for having caught this error in the sim. I got assigned to my first mission shortly thereafter. There may not be a connection, but one thing is certain: aiming to be a zero didn't hurt my chances.

* * *

Approaching a space station, you turn your mind to the technicalities of rendezvous and docking. It's not like parking a car. It isn't intuitive, because orbital mechanics aren't like anything on Earth. When you throw a ball or roll it down a hill, you can predict with fair accuracy where it will go, and how its trajectory would change if you threw harder or softer. But in space, you have to go faster to reach a higher orbit—and once you get there, you actually go slower. So in order to maneuver to link up with another orbiting vehicle, you have to think in a whole new way about how objects behave. Yes, you've got all kinds of sensors and lasers to help you gauge distance and angles, but first you have to understand what they're telling you—and *not* telling you—and how to use them.

My first space flight, in 1995, was all about rendezvous and docking, since the purpose of our mission was to add a permanent docking module to Mir so the Shuttle could go back and forth regularly. Just a few years earlier, I'd been intercepting Soviet bombers for NORAD, but now I was part of a mission aiming to help create a closer relationship between the United States and Russia. When the U.S.S.R. dissolved in 1991, its space program was in danger of dissolving, too, as government funding evaporated. The U.S. didn't want Russian military technology being sold off to or shared with politically unstable countries, so NASA did what it could to shore up Roscosmos, its Russian counterpart, by providing funding for cooperative ventures such as regular visits to Mir. Of course, there was something in it for NASA: learning from the people who had the most experience building and maintaining space stations, and in the process creating a partnership that is absolutely vital today. Now that the Shuttle is not in service, we couldn't get up to the ISS without the Russians. Ultimately, it was a very smart move for both countries to figure out how to work together on space exploration.

But in November 1995, the linkage of the two space programs was still a work in progress. The Shuttle had managed to dock with Mir only once before, earlier that year, and that had involved reassembling an entire module of the space station in order to jerry-rig a spot. It was not a viable or safe option going forward. Which is where we came in: our job was to build a permanent dock. The docking module—which looked a lot like a giant version of a propane container you might hook up to your barbecue, only orange—had been assembled on Earth and then placed in *Atlantis*'s payload bay. Once we were in space, we had to attach the module securely to the top of our craft, then ease on up to Mir and connect. Which we really, really hoped would work, given that it had never been tried before. Since the Shuttle flew rather awkwardly at best, docking promised to be a form of elephantine ballet.

My role in all this was to operate the Canadarm, the Shuttle's robotic arm and the crown jewel of the Canadian Space Agency. I knew it was a national treasure, but to me it was a tool, like a hammer or a farm implement. I would use it to reach into the payload bay, carefully haul the docking module out into space, rotate the module to a vertical position, and maneuver it to within a few inches of our docking mechanism. In order to get them to connect, we then had to fire all the Shuttle's maneuvering thrusters and slam into the docking module, like two trains coupling. If we did it correctly, hooks and latches would engage to form a solid, airtight seal. If not, well . . .

I'd been practicing robotically lifting, turning and manipulating large objects for a full year beforehand on Earth, but of course we were worried—really worried—that plan A might not work. So we had a few backup plans. If firing the Shuttle thrusters to drive us up into the docking module didn't work, we'd try to slam the module down into place using the Canadarm. Since the arm is like a large pair of forceps, designed for positioning things, not for ramming them, there was a chance it might break if we tried this, in which case the 5-ton module would float off serenely into outer space.

Helping to lose a docking module on my first space flight would put me well below minus one, so I really wanted plan A to work. Thankfully, it did. By the end of the second day of our mission, we had what looked like a huge tower sticking out of the top of the Shuttle. Now we had to dock with Mir, which looked like a thick pole with spokes radiating out of it. One drawback of our new 15-foot tower of a docking module is that it blocked any view of where we needed to ease in, exactly. On Earth we'd rigged up a simulator to practice, of course, and had figured out that the camera on the elbow of the robotic arm would be the same height as the place where we'd need to link with Mir. Yes, the angle would be weird, but at least it would give us a visual.

As it happened, that one camera turned out to be crucial, because when the time came to dock, all of our distance sensors malfunctioned. Every single one. They were lying to us, basically, giving us the wrong information about range and speed, so we had no choice but to try to dock by eyeballing it, via the camera view. Fortunately, we had a good idea how to do that, because our instructors had insisted that we memorize every sensor reading from rendezvous to docking, which seemed ridiculously theoretical at the time but meant that we did have a good idea of how to do this manually.

Nevertheless, as you might imagine, there were some very tense minutes while Ken Cameron, our commander, got *Atlantis* into position. If we came in too tentatively, we'd bounce off and have to wait 24 hours to try again, because we needed to attempt docking while we were over Russia, so that the crew on the station could communicate with Mission Control in Korolev (Mir did not have continuous communication with the ground). During that 24 hours, we'd be using up fuel and running the risk of something else breaking, plus we'd still be facing the same problem when we tried again—at which point we'd also be risking total mission failure. However, if we came in too fast and too aggressively, we might collide with the station and cause it to depressurize, in which case everyone inside would be dead in a matter of minutes.

Ken opted not to over-control or under-control. He aimed to be a zero, just relied on his training and wisely didn't try to add any flourishes now that we had a giant barnacle clinging to the top of our vehicle. It worked. We wound up linking up with the docking module just three seconds early. Perhaps you can imagine our sense of relief and anticipation when the moment finally came to open the hatch and enter Mir. Cue triumphant *Chariots of Fire*–type theme music, appropriate to an historic moment of international cooperation.

Only, we couldn't get the hatch open. On the other side, they were kicking it with all their might. But the Russian engineers had taped, strapped and sealed our docking module's hatch just a little too enthusiastically, with multiple layers. So we did the true space-age thing: we broke into Mir using a Swiss Army knife. Never leave the planet without one.

As we floated into the station to greet the waiting crew—Russians consider it unlucky to shake hands across the threshold, and wanted us to wait until we were all the way inside—there was a magical, faint tinkling of bells and chimes. It took me a moment to figure out that this was actually the gentle clanging of old experiments that had been tethered to the sides of the metal spaceship, awaiting disposal or return to Earth.

While we were still in transit we'd resolved to be good houseguests: to help out with the chores, not get in the way, and bring gifts (including a specially made collapsible guitar, called a SoloEtte, which I got to play one evening in a two-crew, three-nation singalong). The not-getting-in-the-way part turned out to be the most difficult. The station was so cluttered that navigation required extra care; to get from one section to another, we had to pull ourselves through narrow, circuitous tubes that were like flexible ventilation ducts. It was a strange feeling, like being inside the intestines of a giant but not unfriendly robot, and in our few days there I learned to do it quickly, so I'd pop out the other side and the rush of air would set the experiments to chime-like tinkling again.

When we got back to Earth, a lot of people asked whether everything had gone the way we'd planned. The truth is that nothing went as we'd planned, but everything was within the scope of what we prepared for. That was one of the fundamental lessons of STS-74: don't assume you know everything, and try to be ready for anything. The other lesson, for me anyway, was that when you're a rookie, aiming to be a zero is a good game plan. My goals had been modest—fulfill my responsibilities to the best of my ability, and not distract or cause any trouble for anyone else on the crew—and I'd achieved them.

When you're the least experienced person in the room, it's not the time to show off. You don't yet know what you don't know—and regardless of your abilities, your experience and your level of authority, there will definitely be something you don't know.

* * *

In 2001, shortly after we docked on my second mission—which was my first visit to the ISS—the main computers that ran the ISS failed. All had an inherent flaw and started overwriting their own hard drives. This meant that for all intents and purposes, the Station was dead: it couldn't control its attitude, point its antennae, run its own diagnostics—all kinds of capabilities were gone, and the ground could barely communicate with us. If we hadn't had the Shuttle docked and ready to control the entire combined structure, we would have been in serious trouble. Fortunately, we could use the Shuttle's communications and thruster systems, and we still had oxygen, food and water, so the crew's attitude was just to keep working the problem.

However, because the computers were down, most of the things we'd been scheduled to do were no longer feasible and we wound up with a day when we were at loose ends. My crewmate Scott Parazynski and I were both rookies on Station and given our limited skill set, there wasn't a whole lot we could do to help solve the computer problem. So we went to Yuri Usachev, the commander at the time, and asked, "What's the most useful thing we could be doing right now?" He said he'd really appreciate an inventory of the inside of every single locker in the Russian cargo block. It's a pretty big module lined with cupboards, so we just started at one end and went through every single one, cataloguing all the stuff that was in there. It was a lot like organizing your closets: useful but time-consuming and glory-free. It took us a few hours and was clearly the kind of task two crew members would never have been able to fit into the schedule if the ISS had been fully operational. We joked around and tried to make it fun while we were doing it, and when we finished, we felt celebratory. We'd managed to add a bit of value on a day when otherwise we wouldn't have accomplished much at all.

Later on that same flight, after the computer problems had been fixed, I had a similar sort of opportunity. We'd set up a video camera for a media event, but the video feed wasn't making it back to the ground. Someone was going to have to start at one end, untangle all the cables, and test each one. I thought, "Might as well be me." Sure enough, it turned out that although we'd checked them before launch, two out of the three cables were bad, so I scrambled around for others, cobbled something together, threw the switch and got the video up and running. It seems trivial maybe, being the cable guy, but I felt good that I'd worked the problem so we could deliver what we'd promised.

In a way, it feels wrong even mentioning it—I didn't, at the time—because I know everyone else on board did similarly unheralded, unobtrusively helpful things. We've all fixed the toilet in space (it breaks down regularly). We've all wiped jam off the walls (it has a way of floating off your toast and splattering everywhere). On the ISS you have to be ready, willing and eager to do every job, from the highest-visibility stuff right down to rewiring an antenna, because there's nobody else to do it.

But if you are confident in your abilities and sense of self, it's not nearly as important to you whether you're steering the ship or pulling on an oar. Your ego isn't threatened because you've been asked to clean out a closet or unpack someone else's socks. In fact, you might actually enjoy doing it if you believe that everything you're doing contributes to the mission in some way.

Still, I'm human. I like recognition and I like feeling that others consider me a plus one. Which is why, as we approached the ISS on December 21, 2012, I consciously reminded myself to aim to be a zero once we got inside. Back home, it was a big deal that I was going to be the first Canadian commander of the ISS. Up here, there already was someone in charge: Kevin Ford, who would continue as commander until he left 10 weeks later and handed over to me. He and his crew were completely acclimated and had been successfully running the ISS for several weeks by the time we showed up.

My smartest strategy was simply to try not to mess anything up or make things worse. I was sure that once in a while, I'd be able to do something good and make an authoritative decision, but it didn't need to happen in the first hour or even the first week. If I barged in, intent on making my mark, I probably would—just not in the way I wanted.

Two decades into my career as an astronaut, I felt as close to being a plus one as I ever had. And I knew that my best bet of getting the crew to see me that way was to keep on doing what has always worked for me: aiming to be a zero.

10 LIFE OFF EARTH

THE ISS IS A ONE-MILLION-POUND SPACESHIP that's the size of a football field, including the end zones, and boasts a full acre of solar panels. Inside, there's more living space than you'd have in a five-bedroom home. It's so big, with so many discrete modules, that it's possible to go nearly a full day without seeing another crewmate. It's an awe-inspiring international project, this mammoth co-op in the sky, and when we docked there on December 21, 2012, the mood inside our humble rocket ship was one of anticipatory excitement. Every potential obstacle had been overcome. We were eager to tumble out, unbathed and hungry, to stretch our limbs and explore our impressive new home.

Not so fast. Opening the hatch always takes longer than anyone would want: two and a half hours, in our case, because first we had to ascertain that the impact of docking hadn't damaged the Soyuz. It had bumped into the Station with reasonable force and speed; we needed to check all our seals to ensure there wasn't a slow leak. Only when we knew the vehicle was intact could we change out of our Sokhols and into regular blue spacesuits, which, like all Russian space clothing, have straps that go under your feet to pull the pant legs down. That's helpful in zero gravity, where there's nothing to prevent the hem of your pants from migrating well north of your ankle. Finally, we were ready.

The Russians view the opening of the hatch, not launch or docking, as the start of an expedition, and certainly it's true that the moment you float into a space station, you enter a new phase of life off Earth. We'd been tapping on our hatch and the Station crew had been tapping back in response—a comforting sound this far from our planet—but we couldn't see them until Roman clunked our detachable hatch handle into place, turned it until it clicked and pulled down. The hatch creaked open like the door of a haunted house, and then we could see them: cosmonauts Oleg Novitskiy and Evgeny Tarelkin and astronaut Kevin Ford, all beaming and looking much cleaner-shaven than we were.

We emerged to join the rest of the Expedition 34 crew in Rassvet, a long, tunnel-like structure jutting out from the Russian segment of the ISS. This mini-module is narrow enough that you have to float down it single file, which made for an awkward six-person photo op as we bumped and twisted around to face the camera that had been set up to record this moment for posterity. But our smiles weren't forced; we were delighted to be together in this remote place. I knew the crew well, especially Oleg, a former Director of Operations for Roscosmos in Houston, but there was no time to get caught up. Already, there were things to do.

We floated out from Rassvet and into the main core of the Russian segment for the televised post-docking press conference, which was also our first chance to speak to our families since launch—a public private event, complete with reporters. Our families were in Mission Control in Korolev, sitting on a balcony overlooking the flight controllers; they could see the video feed of us grinning at the camera, but we could not see them. Nevertheless, it was wonderful to hear their voices as they took turns at the microphone to tell us they loved us. A few even went so far as to say they missed us already. It was a bit self-conscious and stilted on both sides, this televised intimacy, but it felt good to be able to reassure them that we were fine. My crewmates' children asked their fathers to demonstrate somersaults in zero gravity, and Tom and Roman happily, though probably slightly queasily, obliged. But the biggest laugh of the event belonged to Kyle, my 30-year-old son, who took the mic and deadpanned, "Hi Dad, great to see you launch. Now can I have a pony?" There was only one possible answer and I gave it: "Ask your mother."

Afterward, we had a perfunctory safety briefing, then at last Roman, Tom and I could get our bearings. Roman had the easiest time of it, because he'd lived on the ISS for six months in 2009. Long-duration space travel is in his blood: his father, Yuri, is a highly-decorated cosmonaut who spent 430 days in space, first on Salyut 6, then Mir. Like Roman, Tom had also been on the ISS in 2009, during a 15-day Shuttle mission. In the interim more modules had been added, but both men had a better sense of the place than I did because when I'd briefly visited back in 2001, the ISS had been a construction site, very much a spaceship-in-the-making.

Now a huge, humming, functioning laboratory, the ISS is anything but open-concept; it's not possible to take in the whole interior at a glance. The main structure is a long series of connected cylinders and spheres, only they're square inside, not circular. At certain angles, it's possible to see clear from one end to the other, but poking out along the length of it, like branches on a massive tree, are three Russian modules and three American ones, along with a European and a Japanese module. As you approach each one and pull yourself through the hatch, there's an *Alice in Wonderland* moment where you pause to decide which way will be "up"—it's subjective, no longer dependent on the law of gravity but rather on what you're planning to do next. In Node 3, for example, the treadmill sticks out from the wall, the toilet and exercise machine are on the floor, and to reach the Cupola you float upside down. The whole module is the size of a city bus, so at any point you can have four people in there doing different things, each with a different understanding of which way up is.

Although the Station had grown dramatically in size since I'd last been there, I was surprised to realize, shortly after docking, that I actually had a pretty good idea where everything was—the 3-D sims back on Earth had been extremely accurate. And in some other respects, too, the place felt familiar. The smell, for instance, was instantly recognizable: clean, like a tidy laboratory, with a hint of machine. In the Russian segment there was something else, a whiff of subtle glue-y, wood-shop fragrance. There's a lot of adhesive in there because the walls are just about completely covered with Velcro. In space, if you don't hang on to them, things like spoons, pencils, scissors and test tubes simply drift away, only to turn up a week later, clinging to the filter covering an air intake duct. That's why there's Velcro on the back of just about every imaginable item: so it will stay put on a Velcro wall.

On the ISS, there's never any doubt about whether you're in the U.S. Orbital Segment (USOS) or the Russian segment. The latter is smaller in diameter—spread your arms and you can easily touch both sides—and the Velcro is predominantly various shades of green, which creates a not displeasing submarine-like ambiance. Being in the American segment feels different. When the first piece of it—Node 1 (Unity)—was launched in 1998, the psychiatrists who were consulted thought that soothing colors were the key to mental health, so they chose . . . salmon. Either they changed their minds or stopped dabbling in interior design, because the rest of the USOS is, mercifully, white. NASA views too much Velcro as a fire risk, so there's less of it there, and most of it is off-white. Even though the cylindrical segment is 15 feet in diameter, the racks that have been installed to hold experiments and create storage space reduce the interior to a square cross-section where, arms outstretched, it's not quite possible to touch both sides. The combination of bright lighting, no windows and white walls creates an atmosphere similar to that of a hospital corridor.

It's noisy like a hospital, too. Without gravity, heat doesn't rise, so air doesn't mix and move; the fans and pumps that are necessary for comfort and survival whir, clunk and hum, a continuous blur of sound that's occasionally punctuated by the loud ping or bang of a micrometeorite hitting the Station. (Armor protects the ISS from micrometeorites, and while we're sleeping, metal shutters cover the windows for added safety, but none of that would be much use against a big meteorite—you'd just have to scramble into your Soyuz and hope for the best.)

That first day, we were still adapting to a new time zone—the ISS is on Greenwich Mean Time—and by 11:00 p.m., I was definitely ready to call it a night. The six sleep stations spread out between the USOS and the Russian segment are far from luxurious, but compared to the out-in-the-open sleeping arrangements on the Shuttle and Soyuz, they are cozy retreats and, though not soundproof, the quietest places on board. Each one is a white, padded, totally private container about the size of a phone booth, complete with a door and a sleeping bag tethered to one wall. On the other walls are elastic straps (I used them to trap a book, a change of clothes and a small bag of toiletries) and spots for two laptop computers, one solely for work and one for personal use. Velcro on the ceiling helps secure small items like nail clippers and a Sharpie—the preferred writing utensil on orbit since you can hold it any which way and it still works.

In zero gravity, there's no need for a mattress or pillow; you already feel like you're resting on a cloud, perfectly supported, so there's no tossing and turning to find a more comfortable position. Once in my pajamas (Russian-made, long john–esque) I zipped myself into my hooded sleeping bag, which resembled a cocoon with armholes. From my Shuttle days, I knew that a dormant astronaut is an interesting sight, with both arms floating in front Frankenstein-style, hair fanned out like a mane and a facial expression of utter contentment. Turning off my little light, I was perfectly at ease in this otherworldly place, knowing that in Houston and Korolev, people in Mission Control were keeping watch as we spun through the sky and into sleep, on our journey around and around the world.

* * *

Although the ISS is all about cutting-edge technology, living there is in some respects the ultimate off-the-grid experience. It's remote all right, and there's no running water—without gravity, it would cohere into blobs, float away and wreck the sophisticated equipment that keeps the Station going.

The rough-and-ready, improvisational quality to life on board is reminiscent of a long trip in a sailboat: privacy and fresh produce are in short supply, hygiene is basic, and a fair amount of the crew's time is spent just on maintaining and repairing the craft. And there's another similarity, too: it takes us a while to get our sea legs.

Weightlessness doesn't feel the same on a huge spaceship where you can move around freely as it does on a tiny rocket ship where there's nowhere to go. Imagine floating in a pool without water, if you can, then endow yourself with a few superpowers:

you can move huge objects with the flick of a wrist, hang upside down from the ceiling like a bat, tumble through the air like an Olympic gymnast. You can *fly*. And all of it is effortless.

But effortlessness takes some getting used to. My body and brain were so accustomed to resisting gravity that when there was no longer anything to resist, I clumsily, sometimes comically, overdid things. Two weeks in, I finally had moments approaching grace, where I made my way through the Station feeling like an ape swinging from vine to vine. But invariably, just as I was marveling at my own agility, I'd miss a handrail and crash into a wall. It took six weeks until I felt like a true spaceling and movement became almost unconscious; deep in conversation with a crewmate, I'd suddenly realize that we'd drifted clear across a module, much as you might gently bob around in a pool without really noticing.

The absence of gravity alters the texture of daily life because it affects almost everything we do. Toothbrushing, for instance: you need to swallow the toothpaste—spitting is a very bad idea without the force of gravity or any running water to help stuff go down the drain and stay there. Hand washing requires a bag of water that has already been mixed with a bit of no-rinse soap; squirt a bubble of the stuff out through a straw, catch it and rub it all over your hands—carefully, so it clings to your fingers like gel instead of breaking into tiny droplets that fly all over the place—then towel dry. Long, hot showers are out, obviously. Of all creature comforts, they were what I missed most; a wipe-down with a clammy cloth is a poor substitute. Hair washing involves scrubbing your scalp vigorously with no-rinse shampoo, then drying off carefully to be sure stray wet hairs don't wind up floating all over the spacecraft and clogging up air filters or getting in people's eyes and noses. The shampoo worked, more or less, but my hair and scalp never felt the way they do on Earth.

There's no such thing as no-rinse laundry soap, so even ineffectual cleaning of our clothes was impossible. Instead, we just wore them over and over, until they wore out. I'd never been on a long-duration mission before, and I will admit that I was a little concerned about the olfactory implications. Would life in space, um . . . stink? The answer, surprisingly, was no. Admittedly, my sinuses were mildly clogged throughout—without gravity, fluids accumulate in your head—but I never once smelled body odor on the ISS. The reason, I think, is that your clothes are never really in contact with your body; they sort of float next to you, loosely—and given how little we exert ourselves, I'm sure we sweat less, too. A pair of socks lasted me a week, a shirt was good for two weeks, and shorts and long pants could be worn for a month without unpleasant social consequences. When I thought I couldn't get one more wear out of something, I'd cram it into one of the waste containers destined for a Progress, the Russian resupply vehicle that delivers cargo to the Station and then burns up on its way back to Earth.

I went through gym clothes faster than anything else, replacing them about once a week. Exercise is mandatory during a long-duration flight: we'd waste away, literally, if we didn't do it. We have to work out two hours a day to keep our muscles and bones strong enough to handle the extreme physical demands of spacewalking and also to ensure that when we do get back to Earth, we are still able to stand on our own two feet.

Getting exercise isn't all that easy in an environment where movement is so easy, though. It requires special equipment: a stationary bike we clip our shoes into so we don't float away, and a treadmill with a harness contraption that pulls us down so we run on the moving track rather than through thin air. I started with a load that was about 60 percent of my body weight, but the longer

I was in space, the more I increased the load to make the workout more challenging. I can't say that running is my favorite thing to do in space: after you get used to floating everywhere, it feels odd and a little unfair to have to move your legs to go nowhere. Having a hockey game or movie to watch on a laptop while I ran sure helped. (Astronauts who are serious runners seem to mind less; in 2007, Suni Williams ran the Boston Marathon in space, which took her only 4 hours and 24 minutes.)

I also did regular sessions on an Advanced Resistive Exercise Device (ARED), an ingenious machine that uses vacuum cylinders to apply a load of up to 600 pounds to a bar or cable, so that we have to lift against the suction. It was a lot like weight lifting in terms of both the sensation and the physical benefits, and I also used the ARED to do heel lifts, squats and other exercises that would be far too easy, otherwise. All the equipment on the ISS has vibration isolation systems; some pieces even have stabilizing gyroscopes so we don't wind up shaking or rattling scientific experiments while we're working out.

We also have to be careful about perspiration. When there's no force pulling sweat downward, it just accumulates on your body like a slowly expanding liquid shield. If you turn your head quickly, that huge, wet glob of sweat might dislodge, sail across the module and smack an unsuspecting crewmate in the face. Proper etiquette on the ISS is to have a towel tucked into your clothes or floating beside you while you work out, to soak up your sweat. Later, you hang the towel on a clip so the moisture is absorbed back into the air and, along with urine, can be recycled as water.

Yes, water. Drinking water, actually. Until 2010, water on the ISS came in large, lined duffel bags delivered by the Shuttle or resupply vehicles, but now an onboard purification system helps us reclaim about 1,600 gallons a year. Using filters and a distiller that spins to create artificial gravity and move waste water along, we're able to turn sweat, water we've washed with and even our own pee into drinking water. That may sound disgusting (and I'll admit that I didn't like to dwell on the pee part while enjoying a tall, cold pouch of water) but the water on Station is actually more pure than the stuff that comes out of the tap in most North American homes. And it tastes exactly like . . . water.

Shortly after we got to the ISS, I started making brief videos about these only-in-space aspects of everyday life, which the CSA posted on its website as well as on YouTube. Making the videos was easy for me—I'd just press "record" on an HD video camera and demonstrate something, such as how to use the treadmill or wash your hands. It was more time-consuming for the CSA editor on the ground, who added fun, space-y music and graphics, but the effort was worth it: some of the videos went viral and were viewed millions of times. It turns out that people are genuinely interested in the ins and outs of, say, space haircuts (a crewmate does the deed, armed with an electric buzzer, a.k.a. a Flowbee, attached to a vacuum cleaner that catches all the little bits).

The CSA recognized that we had a golden opportunity to generate interest in the space program, and we put together more than 100 videos while I was on orbit. Educational outreach is part of an astronaut's job, but it's a particular passion of mine. For 20 years I'd been speaking about the space program in tiny town halls, elementary schools and Rotary Clubs—anywhere that would have me, basically. In 2010 I set up a program called "On the Lunch Pad," where I talked with school kids via Skype during my lunchtime.

I have found it frustrating at times that so few people know what the space program does and, as a result, are unaware that they benefit from it. Many people object to "wasting money in space" yet have no idea how much is actually spent on space exploration. The CSA's budget, for instance, is less than the amount Canadians spend on Halloween candy every year, and most of it goes toward things like developing telecommunications satellites and radar systems to provide data for weather and air quality forecasts, environmental monitoring and climate change studies. Similarly, NASA's budget is not spent in space but right here on Earth, where it's invested in American businesses and universities, and where it also pays dividends, creating new jobs, new technologies and even whole new industries.

The motive could not be more ambitious: exploring our solar system, discovering what else is out there. The desire to explore is in our DNA. It's what humans have been doing since the first dissatisfied teenager left the family cave to see what was over the next hill. Most people believe it's worthwhile to discover, as we have in the past 10 years, that 2,000 planets are revolving around other stars in our galaxy. Currently, vehicles are driving around on other planets to find out more about them, orbiters are circling almost every planet in our solar system and robotic probes are expanding our understanding of our own atmosphere and the magnetic field that protects Earth from radiation.

These were the kinds of things I explained when I did outreach work, but I'd learned that before you can persuade people that the space program is a good investment, you have to get their attention. Suddenly, on orbit, that was much easier: thanks to the Internet, we could *show* people what it's like to be in space, in real time. They not only paid attention, our expedition became a social media sensation. The reason is simple: people are inherently interested in other people. They care about the big picture, yes, but they're enthralled by the human aspects of space exploration, the minutiae of daily life on board the ISS. Understandably, then, the most popular videos we made were the ones about everyday space oddities.

Luckily, there was no shortage of them. For instance, after a few months, the soles of my feet were nearly as smooth as a baby's and free of calluses—they only bore weight when I ran. Meanwhile the tops of my feet had become callused from rubbing against the footholds that prevented me from floating off while conducting an experiment, say, or taking a photograph. I noticed, too, that my eyes stung slightly, because the moisture that is normally dealt with by gravity simply sat there on my eyeballs; those hard little bits of sleep that I used to wipe away only in the morning built up during the day, too, sometimes threatening to stick my eyes shut, so I blinked a fair amount.

I think one reason people like hearing about these sorts of things is that it helps them see the world slightly differently, perhaps even with a sense of wonder. On Earth, it's just a given that if you put a fork on the table, it will stay there. But remove that one variable, gravity, and everything changes. Forks waft away; people sleep on air. Eating, jumping, drinking from a cup—things you've known how to do since you were a toddler suddenly become magical or tricky or endlessly entertaining, and sometimes all three at once. People like being reminded that the impossible really is possible, I think, and I was happy to be able to remind them.

What we do in space is serious, yes, but it's also incredibly fun. It's not just about the epic EVA but the M&Ms dancing merrily inside the package, colliding colorfully in weightlessness. Life is full of so many small, unexpected pleasures, not just in space but right here on Earth, and I think I see them more clearly now than I used to because microgravity insists you pay attention. Weightlessness is like a new toy you get to unwrap every day, again and again—and it's a great reminder, too, that you need to savor the small stuff, not just sweat it.

* * *

The first explorers who crossed the ocean in sailing ships didn't blithely set off without considering the practicalities and logistics. Before they ever left land, they tried to figure out which kind of timber would hold up the best and what kinds of food would keep on a long voyage. They tried to reduce the risks and improve the chances of success by thinking through every aspect of the expedition, beforehand.

The ISS, too, is a testing ground, a place to consider the practicalities and logistics of even more ambitious expeditions. We're trying to figure out two things: how to make a spaceship that's fully self-contained so we can safely venture farther into the universe, and how to keep human beings healthy while doing that.

Because of all the exercise we do and because our diet is controlled—no deep-fried food, no alcohol, no sinfully rich cakes and cookies—most of us return to Earth in pretty good shape and with a lower percentage of body fat. But in space, things happen to our bodies that may or may not be bad for our long-term health. When I closed my eyes, for instance, I occasionally saw very faint bursts of light: cosmic rays—high-energy particles from some distant sun racing across the universe and striking my optic nerve like a personal lightning bolt. The flashes were right at the edge of perception, almost as if teasing me to detect them. A lot of astronauts experience this, and it's not particularly bothersome, more just a minor visual event reminding you you're not in Kansas anymore. But, of course, it's related to radiation exposure. On Earth, the atmosphere and magnetic field provide some protection from the radiation of the sun and billions of other stars, but the ISS is constantly bombarded by high-energy particles. So far, there's no evidence that astronauts have a significantly increased risk of cancer or cataracts, but we do absorb more radiation than we would at sea level, and it's worth figuring out what to do about that.

Other anatomical changes associated with long-duration space flight are definitely negative: the immune system weakens, the heart shrinks because it doesn't have to strain against gravity, eyesight tends to degrade, sometimes markedly (no one's exactly sure why yet). The spine lengthens as the little sacs of fluid between the vertebrae expand, and bone mass decreases as the body sheds calcium. Without gravity, we don't need muscle and bone mass to support our own weight, which is what makes life in space so much fun but also so inherently bad for the human body, long-term.

Finding out what causes these kinds of changes and coming up with ways to prevent and counteract them will be important in order to, say, get to Mars—a round trip would take two years at least. Turning up there and not being able to see anything would be a problem. Naturally, the best place to study physical changes related to long-duration space flight is on the ISS itself, so that's an important focus up there.

About half of the scientific experiments our crew did were related to investigating what was happening to our own bodies in space. We ran all kinds of tests to gauge how much our hearts were shrinking, what was happening to our bone density and blood vessels, whether changes were occurring inside our eyes, and so on. We were, to a large extent, lab techs: we didn't interpret data—usually we just collected it. For one experiment, for instance, I'd put a drop in one of my eyes, then Tom would tap my eyeball very gently 10 times with a small pressure gauge called a tonometer; the measurements and images were relayed back to Earth so experts could check out what was happening to the pressure inside my eyeball. Tom and I also did ultrasounds on one another's eyes to get accurate images of the optic nerve, lens and cornea (luckily, I was told later that my eyes are just fine). We also performed several skeletal ultrasounds of one another's spines and hands, remotely guided by experts on the ground, as well as cardiac ultrasounds, which are trickier to do.

It was really gratifying to have reached a skill level where I could get a good image of Tom's heart and know that a scientist on the ground could actually figure out what, if anything, the image meant. Most of the human biological experiments we participated in outlasted our expedition; more astronauts will need to do those same experiments in order to have a scientifically meaningful sample size, and it will be years before we learn the results.

We know before we go to space that we are going to be human guinea pigs, but we are highly informed, consenting human guinea pigs. Scientists and doctors come to NASA to pitch their tests and experiments to us, explaining what they're trying to find out and why, and after these briefings, which take days, we're left with hundreds of pages of information and decisions to make about which experiments to sign up for. Medical scientists will do whatever you allow them to—in the 1990s, crews launched with heart catheters and rectal probes—because there's never enough data and there's never a big enough sample of astronauts to study. I signed up for all the experiments except the ones that required biopsies; I'm willing to inconvenience myself and work hard, but not to give away pieces of my flesh.

Urine, however, is another matter, and all of us spent a great deal of time on Station collecting it. The ISS toilet is located in a white booth and consists of a long hose coming out of the wall with a yellow funnel to pee into, just like a mini-urinal. There are foot and hand holds, so you don't float away; you grab the hose, which is attached to the wall with a bit of Velcro, pop the lid off it and wait for it to start drawing air. There's about a 15-second spin-up cycle, and you want to be sure there's good suction or there will be quite a mess to clean up. Even if you pee directly into the tube though, there will always be a few drops left on the funnel. Tracy Caldwell Dyson, who's been singing with me in Max Q, the all-astronaut band, for more than a decade, left an inspirational message on the wall the last time she was on Station: "Blessed are those who wipe the funnel." There's an impressive range of things you can use for that purpose: tissues, baby wipes, gauze, Russian dry wipes and disinfectant wipes. You put whatever you used in a bag, clean your hands with a baby wipe and stick that in the bag too, pinch it closed, put it in the garbage and you're done.

Unless, that is, you're taking part in an experiment of some sort and peeing for science, as astronauts are about 25 percent of the time. In that case, you need to cart some paraphernalia into the bathroom. If all you're doing is testing pH, to check on organ function and body chemistry balance, it's not so bad. You get yourself set up with a data chart, a color chart, a Q-tip, a pH strip, wet wipes and a little bag—all of which, naturally, are prone to drifting away (for some reason, there is no sim at JSC where you learn to corral a bunch of small, weightless objects while also holding a hose and attempting to relieve yourself). This is where the ingenuity born of decades of sophisticated technical training came into play: After a couple of days I figured out that I could stick all the smaller items into one of the bathroom books, which made a decent little trap. Then, after I was done, I could use the Q-tip to swab a few drops off the funnel, rub the Q-tip on the pH strip, match the strip against the color chart to get a valid reading, enter the data on the chart, then clean up as usual. The first time took 15 minutes, but with practice I was able to get it down to 5 minutes.

Collecting a urine sample was quite a bit more complicated and required a container of test tubes, a whole cleanup kit and a big plastic bag that looked just like a hot water bottle, only at one end there was a condom and, at the other, a long, thin hypodermic-looking tube capped with a blue rubber diaphragm. Already inside the bag was a chemical that needed to be mixed with the urine sample for the whole exercise to work. Full disclosure: I'm not entirely sure how female astronauts go about this, but as will become clear shortly, it's almost certainly different than the way male astronauts do it.

First you need to stretch that hot water bottle–esque bag to be sure the little septum between the condom and the bag is as open as it can be, so the force of your pee will overcome the little one-way valve and fill up the bag rather than squirting back out and all over you, all over the walls, all over—you get the picture. Once the bag is filled, you put it in a Ziploc bag just in case it leaks (at least once, it will) and shake it vigorously to make sure the chemical is mixed well with the urine.

At this point, when your hands are covered with blobs of urine and drops are floating around the bathroom, too, it's usually helpful to remind yourself that you are doing all this in the name of scientific inquiry. Take a minute to clean yourself up and while you're at it, grab a disinfectant wipe—surely you've got a free hand!—and clean the ceilings and walls, too.

All right, it's time to fill the test tubes: depending on the experiment, sometimes you'll only need to fill one, but typically it will be five. With a Sharpie, label each test tube with the time, date and your name. While you were shaking up the urine and chemical, bubbles formed in the sack, so now you need to spin it—gently!—like a centrifuge, so all the bubbles collect at the condom end. Then, through the little blue diaphragm, fill each test tube three-quarters full so there's room for expansion after the sample freezes. Luckily, the tubes have Velcro on them so you can stick them to the wall. Once you're done, seal up the big bag in the Ziploc, burping out any air, and clean yourself up again.

Now it's time to fire up the bar code reader and bar code the test tubes, then put them in a mesh bag and place it in a special -140 degree freezer, called a MELFI. It looks like something you'd see in a morgue, complete with sliding drawers that contain long, rectangular boxes. They're so cold that you have to wear special white gloves to handle them, and you can only keep the freezer open for 60 seconds, so you don't compromise any of the other biological samples already in there. That's tricky, though, because as soon as you open a box, a bunch of previously filled mesh bags come floating out. Like a beekeeper, you've got to shove them back in the hive along with the new bag and close that drawer cleanly—if even a tiny corner of fabric gets caught, the thing will jam. This is actually something we practiced doing on the ground, where, of course, nothing was weightless and trying to escape. Here comes the fun part (seriously): as you slide the drawer back in, it flushes out ice crystals that envelop your upper body like the coolest cloud.

Take off your gloves: you're all done! And the whole procedure only took 40 minutes or so. Now you know how much time you'll need to budget every single time you pee over the next four days, which is typically how long you have to give samples for any one experiment. Oh, and don't forget to coordinate bathroom trips with crewmates who are also urinating for science—the MELFI can only be opened once every 45 minutes.

The science we were doing didn't just involve urine juggling, though. Our crew was also testing a device called Microflow, a toaster-sized box that uses fiber optics and a laser to analyze blood samples and provide readings in less than 10 minutes—a wonderful, portable technology that could be a godsend in rural communities. We also worked on RaDI-N 2, a Canadian experiment to detect and measure the levels of neutron radiation in different parts of the ISS. I liked it because it was both simple and elegant: test tubes filled with clear polymer gel were placed in different locations on Station—when a neutron struck a test tube, it created a visible gas bubble. A reader then analyzed the tubes to determine which modules of the ISS were getting higher doses of radiation. (It turns out that some modules are better shielded than others, though it's not yet clear how big a problem this is or what the long-term health implications are for astronauts and cosmonauts.)

Some of my favorite experiments were the ones attempting to answer really big questions like, What's the universe made of? The Alpha Magnetic Spectrometer, mounted on the Station's exterior, is collecting dark matter and high-energy particles to try to provide an answer. Another experiment is looking at the behavior of nanoparticles and how they coalesce without the weight of gravity. Most of the 130 experiments on board are ones that simply cannot be done on Earth: we're there to make sure that scientists on the ground get the information they need.

It's a big responsibility and an honor to work in that huge orbiting laboratory. Figuring out how to support life in the hostile environment of space has resulted in thousands of down-to-earth spin-offs, from temperature-regulating underwear to heart pumps that rely on Shuttle fuel-pump technology. The concrete benefits and by-products of the science we do in space have touched fields from agriculture to medicine to robotics. Data gathered on the

Shuttle and ISS help power Google Maps; experiments with different dietary and exercise protocols have revealed how to ward off, permanently, one debilitating type of osteoporosis; the robotic machinery now used inside the parts of nuclear power plants that are too hazardous for humans is a direct descendant of Canadarm2—the list goes on and on.

A lot of times the work isn't glamorous, but that's okay. The workplace itself is, after all, in a pretty great location.

* * *

Every morning on the ISS, NASA sent us a schedule of what we were supposed to accomplish, broken down into five-minute increments. Almost every day had the same three components. First, some basic maintenance—checking systems, cleaning up, inspecting equipment for wear and tear, that sort of thing. Sometimes, there were scheduled repairs, like overhauling the communications system. Another part of each day was devoted to science: I'd be assigned to work on X experiment for Y minutes while Tom was busy with Y experiment for X minutes, and so on. Often we were in different modules, working on completely unrelated tasks. And finally, there was downtime.

It was a regimented existence but, in many respects, easier duty than I'd had on Earth. I wasn't constantly on the road; I wasn't endlessly training for contingencies. There was a bit of on board training—practicing robotic skills on a simulator or Canadarm2, conferencing with instructors to prepare for an upcoming vehicle rendezvous—but overall, there were fewer demands on our time, and sometimes we were even able to complete tasks faster than anyone on the ground thought we would.

So what do you do on the ISS if you're 10 minutes ahead of

schedule? Well, you can look out the window—I viewed every spare minute on the ISS as a good opportunity to drink in the view. Another thing we like to do with any unlooked-for free time: take advantage of weightlessness. It was not uncommon to come across a crewmate pirouetting, spinning or flipping around just for the fun of it. We also liked to play with water. Someone would carefully squirt a swirling ball of water out of a drink bag and then, like kids chasing a soap bubble, we'd move around this floating ball and blow on it, gently. If we weren't careful, of course, it would break apart and make a giant mess; the forced airflow that draws objects toward an air inlet made steering the water bubble more challenging, and sometimes the only way to avoid disaster was to quickly slurp it up. A few times we used dental floss to corral and spin the ball, laughing and chasing it until it got too close to a wall and we had to smother it and soak it up with a towel.

If we felt like living dangerously, we'd play this game with a ball of coffee or juice—you risked a messier mess, but the colors were good for arty photographs. We also took pictures of balls of water, trying to capture our own upside-down reflections in them. The pepper that we put on our food is suspended in oil so it doesn't fly all over and cause sneezing fits, and once I very cautiously squirted some pepper oil into a floating water ball, creating a delightful sphere within a sphere, held apart by their natural repulsion.

Another game was created spontaneously during one short break in the schedule. The ground support teams use bubble wrap to pack fragile items for launch, so after we unpacked an experiment we put the bubble wrap in a big duffel bag at the farthest end of the Japanese laboratory, where it wouldn't be in the way. On a semi-regular basis, then, we had to float all the way to a far corner of the Station to deal with leftover bubble wrap. A full-length traverse of the ISS is a natural excuse for testing our prowess as space-movers—efficient elegance is a real source of pride for most astronauts, myself included—and soon we'd made a game of it: Who could fly from the Node 1 dining table to the bubble wrap bag, deposit a scrap of wrap securely and get back the fastest? After a while, we actually started hoarding bubble wrap during the day so we could have timed heats at dinner. Watching each other careen down to the Japanese module, arms and legs akimbo, clutching a small, bubbly square of plastic, then swing madly around the corner only to reappear several seconds later, frantic to fly across the finish line in first place, made us laugh every time. I remember being inordinately proud of completing the trip in 42 seconds one day.

There was scheduled free time, too, at the end of most days, and we had a reduced workload on the weekends. Mindful of the need to provide some leisure activities, the space agencies make sure there are DVDs and books on board. There are also musical instruments: a keyboard, ukulele, didgeridoo and guitar. National pride compels me to report that the guitar is a Larrivée, named after its Vancouver maker, Jean Larrivée. Getting it on Station wasn't as simple as running into the factory and grabbing one: everything we take up has to be tested to ensure it doesn't emit too much electromagnetic radiation and isn't off-gassing chemicals, such as benzene, that would be dangerous to inhale in an enclosed space.

That guitar tested me, too. Weightlessness affected the way I played chords: at first my hand overshot the mark, anticipating resistance where none existed, and missed the frets. It took me a while to get the hang of it. On the plus side, I didn't need a strap; the guitar just hovered in front of me, though I did need to brace it against my body to stop it from escaping altogether. One thing remained the same, though. Music sounded just like it does on Earth, despite the whirs and clunks of the fans and pumps, the creaks and snaps of expanding metal as we went in and out of sunlight. Sometimes the background noise was so loud I felt like I was playing in the back of a bus; it turned out that the best place to make music was my own sleep station. Tom and Roman also play guitar, so most evenings you could hear melodies emanating from one sleep pod or another, like music from a nearby campfire.

A lot of people think it must be lonely on the ISS, so far from Earth. But we had multiple links to the ground, ranging from ham radio to VHF to the Internet; our laptops communicated with a server in Houston via satellite relay, thus we could jump online. We had that data link about half the time; though it was slower than dial-up, and streaming videos severely tested our patience, it was fine for email. Far from feeling out of touch, we made a deliberate effort to stay on top of current events. On the day of the Boston Marathon bombings, for example, I actually knew more about what had happened than the CAPCOM I'd called. There was no shortage of people to talk to on Earth: Mission Control was omnipresent, and family and friends back home were just a phone call away.

In fact, at the beginning of our expedition I was calling my kids once a day, until Kyle finally said, "Dad, why do you keep calling? We get it: you're safe!" Apparently the thrill of a phone call from space had worn off. The two-second delay on the line, that irritating echo, didn't help matters. On Earth, my family doesn't typically talk very much on the phone because the kids are so far-flung, but we do communicate constantly via a family Skype chat room: Kristin is at university in Ireland, Kyle lives in China and Evan was, until recently, at university in Germany.

I couldn't easily access the site on orbit, though, so instead I got into the habit of phoning and emailing with Helene daily, and primarily emailing with Kristin and Evan. Kyle, though, still had to put up with some phone calls because he's not a good emailer. He's a professional poker player, so we'd talk about his results, how he liked Wuhan, the city he'd recently moved to, and what he'd done with friends lately—I wanted to hear about his life, not talk about my own. I was already doing plenty of that via videoconferences with schools and reporters. Kyle has a quick wit and an offbeat view of things, and talking to him always made me feel connected to Earth.

I missed my children, but no more than I do on the ground, where I don't see enough of them either. And I missed Helene, though we actually spoke quite a bit more than we normally do when I'm on the road. But I wasn't lonely. Loneliness, I think, has very little to do with location. It's a state of mind. In the center of every big, bustling city are some of the loneliest people in the world. I've never felt that way in space. If anything, because our whole planet was on display just outside the window, I felt even more aware of and connected to the seven billion other people who call it home.

I felt connected, too, to my crewmates. On the ISS, cosmonauts and astronauts are scheduled separately, and the two segments of the spaceship are separate, so you have to make a deliberate effort to see each other. We did that during our five months there, sometimes just by floating over to hang out together after dinner for 15 minutes or so. Mealtimes are very important opportunities to socialize, especially when there are just three of you on board. After Kevin's crew left, Roman was all alone in the Russian segment, so we encouraged him to come have meals with us whenever he could, and often he, Tom and I would wind up talking afterward and listening to music—Roman had a mind-boggling selection on his iPad.

Preparing meals is not laborious on a space station. All liquids, including coffee and tea, come in pouches; most are powdered, and we simply add water, then sip through a straw. The majority of the food on board is dehydrated, so again, we just inject hot or cold water directly into the packages using a kind of needle, then cut open the packages and dig in. There's a lot of sticky stuff like oatmeal, pudding and cooked spinach, because it clumps and is therefore easier to trap on a spoon and get into our mouths without having to chase it all over the place. We had fresh fruit and vegetables only about once a month, when a resupply vehicle or another Soyuz arrived. Once, we got a fresh, crunchy green apple and an orange apiece. Another time, it was a banana, two tomatoes and two oranges. One time, a whole onion each!

Despite the absence of a refrigerator, which is a limiting factor, space food is, for the most part, tastier than you might expect. There's quite a bit of variety: a mixture of Russian food—beef stew, steamed salmon—and American dishes, plus specialty items from other countries. I also got bonus containers of Canadian treats like smoked salmon, buffalo jerky, a tube of maple syrup—even Tim Hortons coffee, the preferred caffeinated beverage on board (Roman took to calling everything else "deputy coffee"—second-best).

Many astronauts, myself included, crave spicy foods after a while, because the congestion that comes with weightlessness means that things taste pretty much the way they do when you have a head cold. Everything is just a bit more bland. My favorite dish was a bag of shrimp cocktail with horseradish sauce, which not only tasted good but had a kick that helped clear my sinuses.

Sometimes we did get a little ambitious and whip up something special for ourselves, like, say, a peanut butter and jelly sandwich. There's no bread on board—crumbs would be a real problem—so we used specially packaged, mold-resistant tortillas. Other times, we planned a special meal, like a breakfast to celebrate the Russian EVA in April. We collected waffles and maple syrup—strange breakfast items for Russians—Brie cheese, smoothies and dehydrated strawberries. All six of us lingered a long time that Sunday morning, floating around what felt like a windowless rec room, having one cup of non–deputy coffee after another, talking and laughing and feeling we were the luckiest people off Earth.

The fact is that even the least eventful day in space is the stuff of dreams. In some ways, of course, it's the improbability of being there at all that makes the experience so transcendent. But fundamentally, life off Earth is in two important respects not at all unworldly: You can choose to focus on the surprises and pleasures, or the frustrations. And you can choose to appreciate the smallest scraps of experience, the everyday moments, or to value only the grandest, most stirring ones. Ultimately, the real question is whether you want to be happy. I didn't need to leave the planet to find the right answer. But knowing what it was definitely helped me love life off Earth. My main source of frustration, in fact, was that I ever had to sleep. It just seemed like a waste of space, where there was so much more left to do and see and feel.

11

SQUARE ASTRONAUT, ROUND HOLE

When I was 10 years old all I wanted for Christmas was a camera. I loved *National Geographic* and I had this idea: if the astronaut thing didn't pan out, photography would be my fall-back career. I was thrilled when I woke up on Christmas morning and there, under the tree, was a Kodak Instamatic. I lost no time setting up moody shots involving my model car collection and some mirrors, then sent the rolls of film off to be developed. The photos that came back were poorly lit and uninspired. So I took some more. But after spending most of my pocket money to get those developed, I had an epiphany: I was never going to be a professional photographer. My pictures were god-awful. I put the camera away.

Years later, as a wedding present, Helene and I were given a serious camera, a heavy, bulky 35mm Canon, which was almost like lugging around a child. I did learn to take somewhat better pictures by fiddling around with the lenses and settings, but no one would confuse my family photos with art. Once in a while I'd get a good shot, but that had everything to do with luck, not talent.

In space, though, I needed to be able to take decent pictures a little more reliably than that. Fortunately or not, I wasn't the only artistically challenged astronaut, and NASA actually brought in professional photographers to teach us, but it was an uphill battle. Imagine an instructor waxing lyrical about shutter speeds while a bunch of fighter pilots are saying, "Just tell me which button to push again," and you have a fair idea what was going on in the classroom. A few astronauts are extremely talented photographers, like my friend Don Pettit, who knew enough to ask for modified cameras and lenses when he went up to the ISS. His sequential stills of the northern lights created a whole new way of seeing the world. But I was nowhere near that level. When I got to the ISS in 2012, I could point and shoot, but that was about it.

Two years before, the Cupola, an observatory module built by the European Space Agency, had been installed on the Station. From the outside, it looks like a hexagonal wart on the belly of Node 3; from the inside, it is a thing of beauty, a 360-degree dome of windows on the world. There are trapezoidal windows on all six sides and, on the top, directly facing Earth, a round, 31-inch window, the largest ever on a spaceship. It's the ultimate room with a view, but highly functional, too: its command and control workstations let us guide operations outside the Station, including controlling the robotic arm.

To enter the Cupola, you have to scoot past the toilet and exercise machine, as though you're diving to the bottom of a pool, then pull yourself in. Suddenly your whole frame of reference changes: when you look up, you can see the whole world. The Cupola is small, less than 10 feet in diameter at its widest point, and when you're in there, your feet dangle out the end, because it's less than 5 feet high. But none of this matters, because you're inside your own personal planetarium. Visually, it's the closest thing there is to a spacewalk: you can no longer see the

ISS—you've escaped, mentally, and are now surrounded by the grandeur of the universe.

We keep up to eight cameras in the Cupola, which is a photographer's paradise, particularly compared to the small portholes elsewhere on Station. The brilliant orange hues of the Sahara, the blurry smear of smog over Beijing—even I felt the need to pick up a camera and try to capture these sights. My first full day on Station, I grabbed a camera with a 400mm long lens, hoping that someone else had already done the settings, since I didn't really know how, and just started taking photos. It was like looking at the world through a straw: you could fit all of Chicago into a picture but not all of the Great Lakes.

By that point I'd been posting pictures—mostly related to my pre-flight training—on Twitter for two years, primarily because my son Evan had told me to. Evan is the communications guru in our family, savvy about the media in general and social media in particular, and he'd been coaching me for years on new ways to draw attention to the space program. He'd helped me do events on sites like Reddit, where people could and did ask me anything, ranging from technical questions about engines to general questions like whether astronauts are religious (they run the gamut from devout to atheist, but whatever the personal belief system, space flight tends to reinforce it) to personal questions about my greatest fear (something bad happening to any of my children).

Evan's specialty is marketing, and he thought that when I got to the ISS, I should be marketing the beauty and wonder of space. It was my chance to stop telling people how inspiring the space program is, and start showing them. All I had to do was post inspiring photos I'd taken from the ISS. Twitter has the virtues of ease—it takes almost no time to write a few words to accompany a photo or answer a question—and immediacy. I could share the view from the Cupola on Twitter mere moments after seeing it myself.

All this was predicated, of course, on being able to take really good photographs. It was a classic "square astronaut, round hole" dilemma: Evan envisioned me as a messenger of celestial beauty, but when it came to cameras, I was actually Joe Fighter Pilot. I explained this when he visited me in quarantine in Baikonur, and he didn't argue. He was, after all, familiar with my body of work as a family photographer. Actually, he mused, the wording of my tweets could also stand some improvement. They were a tad too formal—"robotic" was, if memory serves, the actual word he used. So what was the solution? He smiled and urged me simply to share the sense of wonder I felt about space.

Fine. I tweeted my first photos from the Cupola on December 22, when I had about 20,000 Twitter followers and, because we had downtime over the holidays, hours to labor over my 140-character tweets. I decided I couldn't go wrong by naming whatever it was I'd photographed, and trying to draw an analogy of some sort, likening rivers to snakes and so forth. Two days later, I tweeted a link to a recording I'd just made of "Jewel in the Night," a song my brother Dave wrote. It was a first take—I'd literally pressed "record" on my iPad and started strumming—so you could hear typical Station noises in the background.

Evan approved, which was an encouraging change, so much so that he decided to do me a favor and post the link on a number of different sites, to see whether it gained traction anywhere. Then he had a brainstorm: Why not record ISS sounds all on their own, with no music? No one who hadn't been there had ever really heard them. So I made a few recordings, which I sent to him as audio files. He posted them on SoundCloud, a social network that has very little crossover with Twitter.

The only way for me to explain what happened next is that my son had some time on his hands over the holidays. For years, he's been an avid player of video games, and this was a game with a purpose: public education. In the meantime, of course, most of the people who work in communications at NASA and the CSA also had holidays, and when they got back to the office in the new year, they were stunned and a little alarmed. On January 2, I had 42,700 Twitter followers; by January 7, there were almost 115,000. Suddenly there were articles in newspapers and magazines as well as online about the photos I'd been posting and my tweets with William Shatner and cool facts about life in space. What was going on?

It wasn't just that my photos were improving, though they were. I was taking 100 pictures a day and starting to develop a better eye. I was learning what to look for: weird colors and textures, discontinuities and surprising shapes, like the island off Turkey that looks, from space, like an exclamation point, or the river in Brazil that looks just like the "S" on Superman's chest. People thought it was cool to see the world through the eyes of an astronaut, and they especially liked seeing what their own regions looked like from my vantage point. The main reason for my sudden popularity, however, was Evan's help on the back end, re-posting things on YouTube, Tumblr, SoundCloud and other sites, and driving more traffic to the photos, the recordings and the CSA videos. To him, it had become a challenge: How many more people could he get hooked on space? In Ireland, Kristin, who is a genius at statistical analysis, was helping him by analyzing, say, the correlation between retweets and new followers (there wasn't one).

My son was a one-man, unpaid band, drumming up excitement about and interest in the space program in a way that made me both proud and grateful. For years my kids had rolled their eyes whenever I launched into a sermon about the importance of public service. But Evan had outed himself: he was a Samaritan in cynic's clothing.

* * *

The media exposure we were getting amazed me, but the handover ceremony on March 14, 2013, when I formally took command of the ISS, wasn't touching because it was televised—it was moving because Kevin Ford made it so. Unbeknownst to me, he'd worked hard on a speech honoring Canada and had arranged to play our national anthem, which showed a real awareness of what this moment meant for a little country.

On a day-to-day basis, being commander wouldn't change my life on Station all that much; if the rest of our time there was uneventful, I might never actually issue a single command. But in a crisis, I'd be ultimately responsible for the safety of the crew and the spacecraft, and that knowledge did change my experience in subtle ways, by creating both a heightened sense of vigilance and a stronger feeling of responsibility for the crew's happiness. For the latter, I relied on a surefire, time-tested strategy: chocolate. On Easter morning, everyone woke up to find a bag of really high-quality chocolate eggs outside their sleep stations, courtesy of Helene, who'd had them shipped up far in advance. I also got in the habit of going to the Russian segment bearing chocolate bars, which met with approval from everyone but Roman, who eyed them longingly while grumbling that he was on a diet.

By now, we had three new Expedition 35 crew members on board: cosmonauts Pavel Vinogradov and Sasha Misurkin, and American astronaut Chris Cassidy. Roman was pleased to have company in the Russian segment after two weeks on his own there, and Chris, a former Navy SEAL who has the work ethic you'd expect with that background, was a welcome addition to the American segment.

We were a happy crew and, not coincidentally, a highly productive one. On Expedition 35, which officially started on March 15, we completed record amounts of science, and yet we still had time to play our bubble wrap racing games. To liven things up, every once in a while one of us would have a videoconference with a famous person. Several years ago, NASA and the other space agencies began organizing these calls to introduce a frisson of social excitement into long-duration flights. Many months before launch, each of us had been asked whether there were any people we'd like to talk to from the ISS. I'd asked for calls with a number of Canadian musicians such as Bryan Adams and Sarah McLachlan; Tom requested a call with Peter Jackson, director of the *Lord of the Rings* films. You'd spend about an hour chatting, long enough to get some real sense of one another's interests and lives.

All of us enjoyed these calls a lot, not least for the surreal thrills they offered. I'll never forget talking to Neil Young, who was in the backseat of his 1959 Lincoln Continental, recently converted into a hybrid; each of us leaned forward, peering curiously into one another's strange vehicles and lives. I asked him for advice on song writing, and he said, "I never write songs, I just write them down," adding that if the song isn't flowing through you of its own accord, it might be a good idea to wait until it is. He also said that he is careful not to judge a song until it's finished, "so that it doesn't get poisoned or stunted." Every time I'm writing a song now, I think about Neil's advice.

As it happened, while I was in space I had a great opportunity to perform a song I had written on Earth with Ed Robertson of the Barenaked Ladies: "I.S.S. (Is Someone Singing?)" We did this for Music Monday, a televised event organized by the Coalition for Music Education, where simultaneously, nearly a million kids all over the world sang along while I sang my part, floating in the Japanese laboratory on the ISS. Coordinating that took a lot of planning, but to have all those kids thinking and singing about the ISS, both inspired and motivated, made every minute of planning worthwhile. I still get a little emotional watching that video, to be honest.

That was one of my high points as commander, in terms of public outreach, though I did dozens of fun Q & A events with school kids around the world. It was never hard to get them excited about the possibilities of space. All I had to do was let go of the microphone so it floated for a few seconds, then answer the inevitable question about how we use the toilet in space, and they were hooked.

Another high point came on April 19, when Roman and Pavel, who would take over as commander when I left, were doing an EVA. As I was heading over to help them, I used the aforementioned toilet. There was the usual reassuring clatter and whir as it started up, then suddenly: nothing. No suction, and a bit of a mess. There was no one to help. Tom was doing an experiment in the European segment, one where he wasn't supposed to move around or even talk on the radio. Chris and Sasha were already busy working on something inside their Soyuz. Because of the way the Station is configured and what everyone was doing, we could no longer even get to the regular Russian toilet.

Ours had to be fixed, pronto, EVA or no EVA. Houston came up with a plan: I had to take out the whole central piece, which has multiple electrical and plumbing connections and pumps nasty stuff, including the chemicals that help treat the waste, which meant I needed to wear gloves, goggles and a mask, and also needed to double- and triple-bag each piece of hardware as I took it apart. Just when I'd got everything disassembled, Mission Control in Korolev called: please go close the hatches. Previous crews had messed this step up by trying to rush it; the hatches have to be perfectly sealed or the pressure checks fail and there's a big delay sending the cosmonauts out. I didn't want Roman and Pavel to worry that I was distracted by plumbing issues, so I hastily removed all the protective gear and whipped down there. Once I got all the hatches closed and as soon as they were outside, I dashed back to toilet duty. It went on like this, back and forth, for three hours.

The whole world wasn't hinging on me, but I had to do things carefully in order for Roman and Pavel to get safely outside, and in order to repair the toilet, though I wouldn't be able to tell if I'd actually managed to do so until it was all back together. When the moment finally came to throw the switch I was delighted to hear a beautiful quiet hum, at which point I realized that the old one, which had clattered like a rattling old truck, had been broken for ages. I don't like being single string, having no one watching to say, "Don't forget X," yet I'd managed to pull off two complicated things that were at odds with each other without messing anything up. There's a real sense of satisfaction that comes from being good at living in space and knowing you can do things efficiently and well.

By that point, I also existed in a parallel universe, one where 681,000 people were following me on Twitter; in total, more than

1.2 million were along for the ride, via various social media sites. There were too many magazine and newspaper articles, TV clips and radio mentions for Evan to track. I was being hailed as a photographer, a poet, even a celebrity. I was aware this was happening, of course, but on orbit, none of it seemed real, nor did it bear much resemblance to my everyday life of sweating the small stuff and fixing toilets.

Evan wanted me to do one more thing: make the first music video in space. He wanted me to sing David Bowie's "Space Oddity." He'd suggested this not all that long after I got to the ISS, and was doing all kinds of work on the ground to make it happen, lining up the right people to help with the editing and so forth.

This video, he assured me, would corner the market on wonder. I wasn't entirely convinced, but if there was one thing I'd learned over the past few months, it was to trust Evan's judgment. He'd understood all along that what people are really interested in is other people; showing the humanity of the ISS is what had captured the popular imagination and driven millions of people to go on to watch the CSA's educational videos.

First, Evan rewrote some of the words of the song. In his version, the astronaut lives, and the Soyuz and Station are both mentioned. Next, I recorded the audio track, using a mic and my iPad. Back on Earth, my musician friend Emm Gryner added the piano underneath my vocal; Joe Corcoran played all the other instruments and produced it. Once they were finished, I re-did my vocal over their instrumental track. All told, between January and February, I did three takes, which required a minimal investment of my time.

Only after we got David Bowie's permission did I film the video, in late April and early May. Using a camera mounted on a flexible arm, I filmed myself floating through different parts of the

Station. But the real magic had to occur on the ground, where seemingly endless details had to be looked after; some people at the CSA worked evenings and weekends, for instance, reviewing video and doing the legwork to get legal approval.

I was pleased with the video, and Evan had worked out a master plan for its release during my final days on the ISS. But once I'd finished my part, I hardly thought about it. I had something else to think about: a crisis was unfolding on my watch.

* * *

More than a year before you go to the ISS, you have to decide as a group which holidays you're going to observe up there. This requires some negotiation because the crews are always multinational. For instance, to the Russians, July 4 is just another day, but most American astronauts expect to have that day off work. On Expedition 35, we'd decided in advance to take off Thursday, May 9, which is a big holiday in Russia: Victory Day, commemorating Germany's surrender in World War II. But a few days beforehand, I asked Houston to give those of us in the American segment a bit of work anyway, because Tom and I would be leaving May 13 and we'd get some downtime on the weekend.

At about 3:30 on May 9, 2013, then, I was puttering around when Pavel came over to say, "There is something interesting you might want to see. Little sparks and fireworks outside." Pavel's English isn't the greatest, so it took me a second to figure out what he was talking about. Then I got it: fireworks, Russia, Victory Day—made sense, though it was surprising that he could see them from space. I floated over to the Russian segment to look out the window: no, it wasn't happening on Earth—it looked like fireflies were coming off the left side of the Station.

Inside, we had no indication of a problem, and my first thought was that we'd been hit by a meteorite and sustained a little damage. Tom took some photos with a big lens and, when we blew up the images, we saw that the fireflies were different shapes, like flecks of paint or little lumps of something. This was unusual and merited a call to the ground, though I had to think for a minute about the wording. "Houston, we have small, unidentified flying objects surrounding the ISS" didn't have quite the right ring to it. I went with something a little more circumspect, telling Mission Control that we were seeing flecks; they agreed with the meteorite damage theory, as they'd seen nothing unusual in our telemetry. We took more photos from different angles, sent them down and went about the rest of our day.

About four hours later, we got word from the ground: the ISS had an ammonia leak on the port side. That's a big deal. Ammonia cools the Station's huge batteries and power conversion systems, as well as the living quarters, via a heat exchanger. There are independent cooling loops, and the one that was leaking cooled a heavily used electrical power bus; without it, there would be a significant Station power-down—we'd be unable to run all the experiments, due to potential overheating or lack of power. I quickly ran through possible options in my head: let the ammonia leak out and lose a critical power string, leave it for the next crew to fix, delay our departure and try to fix it ourselves on short notice—we'd probably need a week to get ready to do a spacewalk. Then, as the hours went by, more bad news: the rate of the leak was increasing. The Station was losing its lifeblood.

Those of us inside it were not in any immediate danger, and in any event, we always had our Soyuz lifeboats to retreat to if things got worse. But as you might imagine, how we were going to deal with this ammonia leak quickly became our sole topic of conversation. Roman, Tom and I were scheduled to undock in less than four days, but how could we? An EVA to try to identify the source of the leak was imperative, and if we departed on schedule, that wouldn't be possible until the next crew arrived from Earth, weeks hence. Pavel and Sasha couldn't do it; they weren't trained to work on the American segment, nor would their Russian spacesuits interface properly with the systems on that part of the Station. And NASA's Chris Cassidy couldn't go out on his own—a solo spacewalk is far too risky.

By 11:00 p.m. the CAPCOM had no news for us, except that everyone at Mission Control was still trying to figure out what to do. So I told the crew we should go to sleep: we needed to be rested and ready for anything the next day. I also suggested to Roman and Tom that they might want to tell their wives that we probably weren't going to be home on time, and then I called and told Helene. She said, "Oh . . . well, so long as you're all right, we'll deal with it." What other choice did we have?

We woke up on Friday at 6:00 a.m., as usual, and first thing, checked our laptops for the daily plan that NASA always sends us overnight. It said, "Welcome to prep for EVA day!" It took me a moment to register this. There'd been no inkling of this the night before, and pulling off an EVA with just one day of prep was unheard of. Usually spacewalks are planned years or at least months in advance; even for unplanned walks, procedures are tested in the pool at JSC first.

But we had no time for that. NASA wanted to conserve as much ammonia as possible, so the plan was to pull out the pump controller box and try to figure out what was going on. When you see water underneath a refrigerator, you don't know whether the leak is from a hose, in the wall or inside the appliance itself—the first step is to pull the refrigerator out from the wall. The same idea was behind this EVA: pull out the big pump box, which is on the very end of the Station, as far as you can go without falling off. And overnight, it had been decided that Chris would be EV1 and Tom, EV2.

In other words, I wasn't going out. I had a moment where I allowed myself to experience the full force of my disappointment. This would have been the heroic climax of my stint as commander: helping to save the ISS by doing an emergency spacewalk. I'd never have another chance to do an EVA—I'd already informed the CSA that I planned to retire shortly after returning to Earth. But Chris and Tom had both done three previous EVAs, two of them together, on the same part of the Station where ammonia was now leaking. They were the obvious people for the job. All this went through my head and heart for a minute or two, then I made a resolution: I was not going to hint that I'd had this pang of envy, or say, even once, that I wished I was doing the EVA. The right call had been made, and I needed to accept it and move on so that we could all focus on the main thing—the only thing, really: working the problem. It wasn't the test I would have chosen, maybe, but it was a test of my fitness to command the ISS. Ultimately, leadership is not about glorious crowning acts. It's about keeping your team focused on a goal and motivated to do their best to achieve it, especially when the stakes are high and the consequences really matter. It is about laying the groundwork for others' success, and then standing back and letting them shine.

It was time for me to do that. It was time to be a commander.

I stuck my head out of my sleep pod and at almost the same moment, Tom and Chris poked their heads up out of their pods: three prairie dogs, all grinning. *Did you see that? We're doing an EVA!* We still thought it was very possible that it would be called off, but we had to get ready. We did the scientific experiments that couldn't be put off, then all three of us focused entirely on preparation. Normally, we'd have had days for that. Now we just had one.

We started working on Tom and Chris's diet, figuring out what they should be eating; they needed lots of carbs, which their bodies would burn more slowly, so they'd have enough energy if they did wind up spacewalking. We had to recharge batteries for the spacesuits, gather all the necessary tethers and equipment, pre-stage the airlock with everything we'd need the next day, resize a spacesuit that had been sized for the next crew, so that Tom could use it—and that was just for starters. Meanwhile, Mission Control was refining the plan. The choreography got more detailed as the day went on and the leak showed no signs of stopping: EV1 would do this, EV2 would do that, and they'd need to have this equipment and those tools. I spent part of the day fashioning something that looked like an oversized dental mirror so they could inspect an enclosed space to look for a leak; using copious amounts of tape and zip ties, I modified an existing mirror to turn it into a spacewalking tool.

Fill the drink bags, polish the visors, get the right number of emergency bottles of oxygen into the airlock, check and double check everything—we needed to be methodical and try to think of everything that could go wrong. One possibility was ammonia contamination: Tom and Chris might be squirted with the stuff when they pulled out the pump controller box, and then we'd have to be sure they were decontaminated before they came back into Station. Ammonia decontamination is a rarely used procedure and one we don't practice much, so I had us do a mini-sim where we looked at all the hardware and worked through the whole matrix of things we might have to do, depending on the level of contamination.

In the meantime, I'd asked NASA to negotiate with Roscosmos to see whether I could get a cosmonaut's help with suit-up and prep the next day. Sasha's English was pretty good, but he was a rookie. Roman had the deepest training in the American spacesuit, but he was packing the Soyuz—a critically important and time-consuming task, because the position of each item affects how the vehicle flies. NASA and Roscosmos both wanted Roman to keep going, so we could undock on Monday. Privately, I thought this was crazy—there was no way we were really going to leave on schedule. Oh yes there was, both space agencies insisted, and they reached a deal: Pavel, who would take over command of the Station when I left, could help.

The next morning, right after breakfast, we got started. I was the intravehicular crew member (IVA), choreographing the suit-up of the spacewalkers and getting everything ready for them to go outside. It turned out to be much more demanding than I'd imagined, and having an extra set of hands was a big help. Pavel is one of those people who, like my dad says, thinks with his hands—he just has a natural, innate sense of how all the fussy mechanisms of space equipment work.

As IVA, there are probably 50 ways to blow it without knowing you've blown it until it's too late, like hooking up a helmet camera improperly. It was clearly an ideal moment to aim to be a zero. My goal wasn't to get Tom and Chris out the door in record time; it was to stick to the procedures, which Pavel and I had never done before, either independently or together. It was finicky, engrossing work and I took huge pleasure in doing it meticulously, in having the language skills to be efficient and safe while tasking Pavel, in making sure our guys, our team, were being set up properly to try to pull off this difficult, dangerous, important job.

Building the spacesuits around the astronauts, getting everything configured right, installing the equipment—it's like assembling a big Meccano robot, and Tom and Chris couldn't help much because they were wearing masks, pre-breathing pure oxygen. The pressure inside the spacesuit is much lower than ambient cabin pressure, so they had to breathe pure oxygen in order to flush the nitrogen from their bodies and ensure they didn't get decompression sickness—the bends. All of this took hours, but eventually we were ready to stuff our crewmates, one at a time, into the actual airlock, then close the hatch and start depressurizing it.

I felt some trepidation. Once you close the hatch to the airlock, you're saying goodbye to redoing anything. I knew I'd been careful, but if I'd messed something up or they were missing a piece of gear, we might not find out until halfway through the EVA. I watched them until they were outside and doing something straightforward, and then I quickly worked through the routine tasks Houston had put on my timeline. But it never left my mind that my crewmates were outside, doing something crucial; I was also very aware of their vulnerability. Relief wouldn't really come until they were back inside.

In the meantime, my role was simply to support in any way I could, so I decided to skip all exercise, just this once. I followed along with the procedures Tom and Chris were doing so I always knew exactly where they were in the process, and I listened to their communications with the ground. When the ISS was out of range of the satellites that let us communicate with Houston, I was ready on the radio with information and the next steps, so Chris and Tom could stay on schedule. Once, as Chris had requested beforehand, I reminded him to say a few words about Marq Gibbs, a longtime lead diver in the Neutral Buoyancy Lab who'd helped us practice spacewalk sims and who'd died in his sleep the previous week, unexpectedly, at just 43. Chris paid him tribute just before they came back into the ISS, pointing out that it takes a cast of thousands to make any EVA possible.

Throughout that five-and-a-half-hour spacewalk, I felt a bit like a choreographer probably does while watching dancers perform; there was a sense of involvement and responsibility, a feeling of shared risk and reward, but also a necessity to detach and trust them to do their jobs properly. It felt good, when they were safely inside the airlock and we were using the ammonia sensors, to be able to say, "Okay, we're going to do what we practiced yesterday." The unknown part was over. It felt even better when it turned out their suits were not contaminated and we didn't have to repeat the entire complicated rigmarole we'd rehearsed.

Best of all, it appeared that they had not only located the problem but fixed it. When they'd pulled out the box that holds the pump module, expecting to see evidence of a leak underneath, there wasn't any, and the box itself was pristine, suggesting the leak was inside it. They'd swapped in a new module, a spare that was stored nearby, bolted it into place, and once they were back inside, Houston gently repressurized the line that circulates the ammonia. No more leak.

When I'd repressurized the airlock and Pavel and I were pulling our crewmates' gloves and helmets off, the feeling was wonderful. We'd beaten the long odds, done our job right, and maybe even fixed the problem and sort of saved the Station. What's more: we were still on track to undock in less than 48 hours.

The crew had come together to pull off an EVA in unprecedented time. The shared feeling of pride was palpable. I was proud of Tom and Chris's hard-won competence, of Pavel's skill even though he was doing something for the first time, of Sasha's willingness to shoulder an extra load so Pavel could help out, of Roman's dogged industry, continuing to pack our Soyuz so we could leave on time.

And I was also proud of living up to NASA's belief that I was capable of commanding the world's spaceship. On my first day at JSC, I hadn't been an obvious candidate. I was a pilot. I didn't have much leadership experience to speak of at all. Worse: I was a Canadian pilot without much leadership experience. Square astronaut, round hole. But somehow, I'd managed to push myself through it, and here was the truly amazing part: along the way, I'd become a good fit. It had only taken 21 years.

PART III

COMING DOWN TO EARTH

12 SOFT LANDINGS

As we were getting ready to leave *Mir* at the end of my first space flight in 1995, the mood was convivial. We were rushing around taking last-minute crew photos, signing sheaves of envelopes (a cosmonaut tradition: Russians are, for whatever reason, avid collectors of envelopes that have been in space) and double-checking that we hadn't left any Shuttle gear behind. As a parting gift, we gave the Mir crew all our remaining condiments, like packages of salsa and mustard, which help make space food taste a little less bland.

I didn't feel let down now that our mission was almost over. I felt that I'd had an experience that no one could ever take away from me—fleeting, yes, but it would be part of me forever, so I was entirely ready to leave. We had done something unprecedented and near-impossible, building a dock for future Shuttle visits, and we'd done it well. As we prepared to undock, there was a palpable sense of triumph inside our spaceship.

I pushed the button to start driving open the hooks that connected *Atlantis* to Mir, and after a couple of minutes, those built-in springs pushed us apart—an effortless kiss-off. As we started to drift away, the ship-to-ship radio crackled to life and the melancholic strains of "Those Were the Days," sung in Russian, filled the Shuttle. We'd all sung the song together the night before in Mir, with Thomas Reiter and me on guitar. At the moment of undocking, the campiness of the song fit our mood perfectly. Spirits were high, as though we'd won a gold medal in the Cosmic Geek Olympics.

We did a fly-around, one perfect looping circle to complete a full photographic survey of the station's exterior. We were (and still are) trying to understand orbital debris—how often it hits spaceships and how big the rocks and dust grains are. Very little orbital debris is man-made; almost all of it is the stuff of the universe, such as meteors and comet tails. Detailed reviews of blown-up versions of these photos, so all the holes and pockmarks could be counted, would provide key data. After 360 degrees of behemoth choreography, with *Atlantis* slowly revolving around Mir like a whale skirting a giant squid, we fired our orbital maneuvering engines, pulled away safely and headed for home. We stayed on the radio, though, chatting and playing a little Tchaikovsky for our friends back on the station, until we lost contact.

The Shuttle was a far more complicated vehicle than the Soyuz, which is highly automated, and landing it was an exceptionally high-demand piloting task. It was very difficult to fly, this hypersonic glider, so NASA chose top-notch test pilots and then trained them for many years to be able to do it right. Simply getting the Shuttle ready to survive re-entry required multiple systems checks and reconfigurations; one trick—we had to point the belly at the sun for hours to warm up the rubber tires for landing. Landing, in other words, required the same degree of focus and preparation as launching.

The lesson for me was that the very last thing you do on a mission is just as important as the first thing you did—perhaps even more important, actually, because now you're tired. It's like the last mile of a marathon: the effort has to be more deliberate and you've got to push yourself, hard, to keep going right to the very end. It's tempting to tell yourself, "I've only got 20 steps left," but if you start anticipating the finish line, chances are that you'll let up and then you could make mistakes—ones that could be fatal in my line of work.

It's dangerous to think of descent as an anticlimax. Instead of looking back longingly over your shoulder at what you're leaving behind, you need to be asking, "What's the next thing that could kill me?"

* * *

I was downstairs on the middeck for that first Shuttle landing, just a hopeful, knowledgeable passenger with no windows, no instruments, no control. My main responsibility was to make sure that everyone on the flight deck was suited up and strapped in. I'd done that perfectly, and was on the middeck alone when Jim Halsell, the pilot, put on his helmet. His communication cord had been floating between the neck ring of the helmet and the neck ring of the suit itself; when the rings locked together they trapped the cord, leaving him unable to talk to our commander or to Mission Control. That's a big problem at any point in flight but particularly when you're trying to re-enter the atmosphere.

I'm not even in my own pumpkin-colored pressure suit yet when Jim hollers, "Come help me." He can't get his helmet open to release the comm cord. On the flight deck, they're doing all sorts of checks and turning on the flight controls, and he's having to yell just to be heard through his big, thick helmet. So I float over to try to pull it off. No luck, the thing is completely jammed.

I need to put more muscle into it, but Jim is belted into a seat that's mere inches below the most critical switches for controlling the vehicle. If I yank too hard and his helmet comes off suddenly, there's a good chance I'm going to smash into that panel and cause a real problem. I pull more vigorously, still wary of the potential for disaster. The helmet doesn't budge.

Picture this, if you can: we're coming down into the upper atmosphere, I'm a rookie still dressed only in my underwear, my stomach's starting to feel queasy and I'm working a problem no one anticipated, while everyone else is fully occupied trying to ensure we arrive alive. Lightbulb: I whip downstairs, find a big, long slot-head screwdriver—the kind you'd use to break open a door—fly back up and try to use it for leverage to unjam the helmet. Meanwhile, Jim is still focused on helping fly this incredibly complicated vehicle, trying to ignore the fact that now my body is wrapped around his helmet to cushion the thing from flying away, and I'm trying to pry it off with the screwdriver, looking, I'm sure, like Bugs Bunny in that episode where he's hugging the head of The Crusher, the monstrous boxing he-man.

Finally, the helmet pops off and I bounce off the ceiling, right myself, untrap Jim's comm cord and refasten his helmet, just in time to drag myself back downstairs and pull on my big orange pressure suit—only, there's a little bit of gravity now, so I keep getting bounced to the floor and I'm starting to feel sick. The suit wasn't really designed for you to put it on by yourself, but it's possible if you work at it, and when I'm finally in, I plunk down in my seat. We're way down in the atmosphere by this point, already Mach 12, I'm sweaty from the exertion and now I realize I've messed up my own comm cord somehow: I can hear what everyone else is saying, but they can't hear me. That's no big loss, as my main focus at this point is trying not to throw up.

I feel like I've only been in my seat for five minutes when we begin our slow, curving turn to line up with the runway in Florida. Since there are no windows I can't see anything, but I sure can hear the rush of air that sounds like a freight train and can feel the very steep final dive to the ground, followed by an elegant touch-down. Our final approach speed is 300 knots, 195 at landing, and then we slow down carefully, thanks to a drag chute and wheel brakes. Only when the motion ceases altogether does the commander issue the radio call: "Wheels stop, Houston."

But the mission was still not really over. We had to refocus and push ourselves physically and emotionally for a last, hour-long burst of effort. There was a 150-step procedure to shut down the Shuttle, and each step was crucially important in order to ensure the vehicle would be ready to fly again in a few months. Only after the ground crew purged the unused toxic, caustic fuels that kept the hydraulic and life support systems running, and covered the fuel nozzles on the front and back of the Shuttle, were we free to exit, unsteadily at best. Some astronauts need to be carried, many vomit, and all of us feel awkward re-adapting to gravity, but an hour later, freshly changed into blue flight suits, we were back to inspect the belly of our spaceship for any damage, greet the ground crew and hold a small press conference.

It was only after all that that I allowed myself to relax. I was a little dazed, but also exhilarated. I'd done my part, and as a crew, we'd fulfilled our mission.

* * *

When we launch from Baikonur, the traditional send-off from the Russian ground crew is, "*Miakoi posadki!*" which means, "Soft landings!" It's a sincere wish but also a joke, because they know very well that there won't be anything soft about our landing when we return to Kazakhstan. Returning to Earth in the Shuttle was a fairly gentle experience, but Soyuz landings are famously rough: high g-forces, heavy vibration, rapid spinning and tumbling, all funneling down to a brutally jarring thud on the unforgiving Kazakh plains.

It's a wild ride, and everyone who's ever taken it seems to have a story about it. My favorite is the one cosmonaut Yuri Malenchenko tells about his return in 2008 with American astronaut Peggy Whitson and South Korean space flight participant Yi So-Yeon. When the Soyuz comes back to Earth, explosive bolts fire so the orbital and service modules are flung away to burn up in the atmosphere; only the re-entry capsule has an ablative shield to protect it from the heat. As Yuri and Peggy's Soyuz started to come back into the atmosphere they heard the explosive bolts fire but, though they didn't know it at the time, one of the modules didn't actually separate from their capsule. It was still attached by one bolt and getting hotter by the second, because as the air got thicker, pressure and friction increased. The re-entry capsule, which wasn't designed to return to Earth with a heavy, burning ball attached to it, became uncontrollable.

As the Soyuz ripped through the sky in pure ballistic mode, the g-force climbed to nine—but it felt much worse than that to the crew because the capsule was tumbling so violently. Instead of just being crushed down in their seats, they were being banged around and squashed every which way. The crew couldn't see what was causing the problem, but they knew something was terribly wrong and that the vehicle couldn't survive that type of punishment much longer.

Fortunately, the aerodynamic forces got so intense that the bolt snapped off, releasing the burning module. But it had hung on so long, at such high heat, that the top of the re-entry capsule was completely scorched. Yuri, who is unusually unflappable, even by cosmonaut standards, felt liquid dripping onto his legs and figured, "Oh, it's molten metal; the Soyuz must be melting." His response was to say nothing, move his legs a bit and continue fighting to control the vehicle (later he figured out that the drips were water from behind an oxygen panel where condensation normally turns to ice during landing). They were seconds away from death, literally.

Then, thanks to its inherently good design, the vehicle stabilized, its parachute actually opened and the crew's capsule subsequently smacked down, very hard but safely, on the ground. But they'd landed well short of the intended target, so nobody was there to meet them. No one on the ground even knew exactly where they were; the fireball of re-entry had disrupted communications for many minutes.

Usually, after a crew has been in space for months, they're too physically debilitated even to open the hatch, so a ground crew is standing by to extricate them. But somehow, after a few minutes, Yuri managed to open the hatch a crack—a superhuman feat given how weak and shaken up he was. Right away he smelled smoke. That was to be expected given the temperature of the vehicle, but when he cracked the hatch a little farther, what he saw was fire, everywhere. The Soyuz had landed in a grassy field and ignited it. By the time Yuri was able to get the hatch closed again, his hands were burning. All three of them wanted nothing more than to get out—they were nauseous and just feeling horrible, sitting in a cramped, now smoke-filled capsule—but the world was on fire. They were in no condition to try to jump out and make a run for it. So they waited. Nobody came.

After a while Yuri decided to risk it and opened the hatch again. Good news: the fire had burned past the vehicle. Somehow he crawled out, and lo and behold, standing there were some locals, a few Kazakh men who'd been drawn by the smoke. They looked at him curiously, and then the only one who spoke any Russian asked, "Where did you come from?" Yuri was trying to explain when the guy interrupted. "Well, what about your boat? Where did the boat come from?" He just couldn't believe that this flat-bottomed craft had really come from space.

In the meantime, Peggy and So-Yeon, whose back had been hurt pretty badly during landing, were working their way out of the capsule, and the guys helped them. At this point Yuri really wanted to get his radio equipment to try to call the rescue helicopters, but he didn't have the strength to go back into the Soyuz and retrieve it. No problem. The smallest guy volunteered, helpfully climbing into the "boat" that had just fallen from the sky and grabbing anything he could lay his hands on. Yuri could see him cramming stuff into his pockets, but he was physically powerless to intervene.

Yuri confronted him verbally, though, and while that was going on, the first helicopter came into sight and promptly radioed back to Mission Control that the capsule had been located but no parachute was visible. It had burned up in the fire, of course, but to everyone who heard that message, it could mean only one thing: the crew was dead. Mass devastation. Quickly followed by mass celebration after the copter landed and radioed back the good news: the crew had survived a ballistic landing, an inferno and some boat-loving bandits.

Though I hoped for a somewhat less eventful ride home, as our time on Station drew to a close I felt a sense of real anticipation about my first Soyuz landing. I'd trained extensively for it and viewed it as a fitting end to my career as an astronaut: a rare experience right on the hairy edge of possible, approached with forethought and a sense of purpose. I've looked forward to every flight I've had as a pilot, but I suspected this would be one of the most memorable of them all.

I was right.

* * *

The last few days of a mission are usually a bit of a blur, because there's so much to do. On top of the regular tasks, we have to practice landing procedures on a computer simulator and pack up our Soyuz meticulously, because where and how each item is stowed affects the vehicle's centre of gravity, which in turn determines how much control we have over it. Typically, the last minute is also when you finally get around to doing all the little things you've been meaning to do for months: shooting a video tour of the ISS to show friends and family back home, taking photos of crewmates in bizarre, only-in-space poses and, just because you can, peeing upside down.

But our mission did not end typically. We had that emergency spacewalk on May 11, a major undertaking just 48 hours before we were scheduled to undock, so everything thereafter was a scramble. Right until the minute we got into our Soyuz, we were flying around—literally—cleaning up the Station, throwing out old clothes and tying up loose ends.

The pell-mell nature of our departure meant that nostalgia had no opportunity to take root, so our Change of Command ceremony on May 12 wasn't momentous or elegiac. It was cheerful and rushed. I handed responsibility for the Station over to the new commander, my good friend Pavel Vinogradov, with a little speech and a big handshake (which didn't work all that well in zero gravity, because our whole bodies moved up and down, so the effect was less than solemn), then got right back to my to-do list.

While Roman focused on the Soyuz, Tom and I did some last-minute science and tried to help set Chris Cassidy up for success. He was going to be all alone in the American module for a few weeks, just as Roman had been all alone in the Russian module after Kevin Ford's crew left. We urged Chris to have dinner with his Russian crewmates, make an effort to socialize and allow himself to enjoy some downtime rather than work round the clock. That evening, Tom, Roman and I finally added our crew patch to the wall. It was number 35 in the long, colorful row, which helped keep sentiment at bay: so many astronauts and cosmonauts before us, and so many yet to come.

At 9:00 GMT that last night, I was reviewing my Soyuz checklists when the "Space Oddity" video was posted on YouTube. I wasn't thinking much about it beyond hoping that it went well for Evan. It had been his idea, his responsibility, his baby, and he was the only person who was nervous about it—a good indicator of ownership. All I'd done was sing, strum and press record. Before I went to bed I quickly checked online to see whether anyone had watched it yet. I was shocked. There had already been close to a million hits.

The very last day on the ISS was a bit like a travel day anywhere. Among other chores, I vacuumed my sleep station and cleared out the few remaining personal items, including my sleeping bag. The next crew would bring new ones; we take ours back with us in the orbital module, in case we have deorbit troubles and wind up having to spend a night or two on the Soyuz. If not, they're jettisoned along with the module and burn up on re-entry. I took a few last photos, cleaned up the Japanese lab, worked a few experiments and reviewed the Soyuz checklists again to make sure I was refreshed.

But despite the flurry of activity I felt a need to steal time, to find a way to be alone in this incredible place, physically and mentally. When I was 7 years old and my family moved from Sarnia to our farm in Milton, I'd had the same impulse. I distinctly remember walking around our Flamingo Drive neighborhood for a last look, fully realizing that my time in that place, which had been a big part of my life and had helped form me, was now at an end. On the ISS I did the same thing. I deliberately went to the Cupola and spent some time trying to soak up the feeling of being there, to internalize what it felt like and what the world looked like from that vantage point. I felt not sad but respectful. I wanted to acknowledge the significance of the time I'd spent on the ISS, and everything it had meant to me.

Then the clock struck 3:30 and, like Cinderella, we were suddenly yanked out of one existence and thrust into another. We said hurried goodbyes to the other crew, tempted to linger with them in that remote place yet knowing we had to stick to the time line. Then we hustled into the Soyuz and closed the hatches. I would not be back in the ISS again, but that was all right. Earth is home to everyone I love.

* * *

Once in the Soyuz, the pace slowed abruptly. It was a dramatic shift, a bit like complete silence after listening to Beethoven's *Fifth* at top volume. We have to do meticulous pressure checks before we trust our hatches, and it takes about two hours before the temperature settles—at first, the Soyuz is chilly—and we can be absolutely sure that we have a tight hermetic seal. The week prior, we'd brought the vehicle out of hibernation and checked the thrusters and motion control system. Since then, Roman had been packing—alone, as only cosmonauts are allowed to pack a Russian vehicle, and under considerable pressure. When Kevin Ford and his crew had returned to Earth, Kevin's seat shock absorber had failed, so he'd experienced a higher g-load, and there was some concern that the issue might have been the way their Soyuz was packed. So Roman had to make sure ours was done just so, and it was.

The re-entry capsule was jammed with medical samples in cold packs and broken hardware that needed fixing—so full, in fact, that we'd had to leave personal belongings on the ISS in "wish to return to Earth" bags. I'd sent a few things back in March, but there were items I'd still needed on board—a favorite shirt, the "recording in session" sign from my sleep station—and now I had to leave them behind and hope they wouldn't remain on orbit permanently. Someday there might be space for them in another vehicle.

One thing I wasn't going to leave behind was my Maple Leafs shirt. After a very long dry spell, the team had qualified for the Stanley Cup play-offs, and tonight was the seventh game of the Eastern Conference quarter-final series. I'd been following it avidly, albeit belatedly, on Station; while running and cycling, I'd watch day-old games the CSA and NASA sent me via data uplink. Leafs fans are stubbornly, some might say irrationally, loyal, not the sort of people who care that they're not supposed to wear team jerseys under their spacesuits. It was May 13, the Leafs were playing the most important game of the season so far—what other choice did I have? I put my shirt on over my long underwear and settled into the left seat. It felt good to be in my spot again in this sturdy little rocket ship.

I was no longer in charge. Roman, our *Commander Soyuza*, was, and he'd flown home in a Soyuz before. Tom and I hadn't, and we also hadn't been in the vehicle for five months, so during the pressure checks we reviewed all the things that could kill us next, talking through what we'd do if the undocking hardware didn't work, for example, and which page we'd turn to if we didn't accelerate properly during the deorbit burn. Roman is a confident, genial leader, and he ran us through the procedures and checks efficiently. Then we started getting into our Sokhols.

They were noticeably more snug. Without the pressure of gravity, the cartilage between the vertebrae in your spine expands and your body lengthens; this was taken into account when our suits were fabricated, but nevertheless, it was surprising to discover at the age of 53 that I'd grown an inch or two. It took each of us about 15 minutes to find a way to scrunch down into our suits, and afterward we closed off the orbital module that had given us a modicum of living space five months earlier, on our way to the ISS. Unless something went wrong and we got stuck in space an extra day, we wouldn't need it; descent only takes three and a half hours. The module was now full of garbage, ready to be jettisoned.

Finally, when we were well cocooned and strapped firmly into our seats with our knees wedged up against our chests, I pushed the command to undock from the ISS. We were on our way.

* * *

Undocking is a peaceful contrast to the fiery pageantry of launch. It takes about three minutes for the giant hooks and catches to release. Our Soyuz is a small barnacle clinging to a massive ship, but gradually little springs push us away and we drift off as our friends watch from the windows of the ISS, waving farewell.

We travel slowly at first, just 4 inches a second, but after three minutes, we fire our engines for 15 seconds and start to pick up speed. Then we coast, relying on orbital mechanics to take us well clear of the Station. We need to get a safe distance from the ISS before lighting our engines again, or the exhaust and spatterings of waste fuel would batter her big solar arrays, in the same way a windstorm batters a ship's sails.

This puts us on a slightly different trajectory than the ISS as we orbit the Earth. Moscow calculates all the new data, such as our deorbit burn time, and we pencil it onto our checklists. It's calm now, but I take anti-nausea meds. I know tranquility is only temporary.

After about two and a half hours it's time: we turn the ship tail-first and set up for deorbit burn, firing the engines for 4 minutes and 20 seconds. There's a critical moment during the burn when there's no turning back—you've decelerated so much that you're committed to falling into the atmosphere. We passed this point and felt the vehicle pushing on our backs, like a solid hand. The sensation is that you're accelerating in the other direction, but actually you're slowing down.

What follows is a wild 54-minute tumble to Earth that feels more or less like 15 explosions followed by a car crash. The Soyuz's trajectory changes from a circle to an ellipse, and when we hurtle down to the low point we begin brushing into the upper atmosphere, where the denser air instantly starts slowing us. It's like sticking your hand out a car window when you're flying down the highway, and feeling the drag of the wind. Then, 28 minutes after firing the engines, the explosive bolts blast open, lobbing the orbital and propulsion modules away to burn up. I think of Yuri, Peggy and So-Yeon, and hope our Soyuz did its job. The loud staccato bangs as the bolts exploded had sounded right, and I saw the fabric that covers the vehicle flash by the window. Then the drag of the air starts to stabilize us and I know we're good. We still have some roll, but there's no way a reluctant module is still hanging onto our capsule.

It's getting hotter and more humid, despite the tough protective hide of the ablative shield. Looking out, I see orange-yellow flames and a stream of high-speed sparks pouring off the vehicle, and hear a series of bangs. Either there's a flaw in the shield or some trapped moisture, or we've got a real problem. I don't say anything, because what is there to say? If the shield fails, we're dead. We are a fiery bullet slicing through space, coming into sunrise.

Two minutes later, at 400,000 feet, the air gets perceptibly thicker. The temperature inside the capsule is still climbing, and my Maple Leafs shirt is drenched with sweat. Now there's even more drag and a rude welcome back to gravity, which squashes us back in our seats. The g-force builds rapidly to 3.8 times Earth weight, which is crushing compared to the weightlessness we've enjoyed for the past five months. I can feel the heaviness of the skin on my face as it's mashed back toward my ears. I take little cheater breaths; my lungs don't want to fight gravity. My arms seem to weigh a ton, and suddenly it's a strain to lift one even a few inches to flick a switch on the control panel. Going from weightlessness to max g and then back to the 1 g experienced on Earth only takes 10 minutes, but it's a long 10 minutes.

Once we've slowed significantly—picture a rock sinking in a deep pond—our drogue chute opens to cut our rate of descent. At 17,000 feet, the main chute opens and we're laughing, yelling, "Yeehaw!" The Soyuz is spinning and whipping around crazily, rattling and twisting too quickly, even, to make us sick. Then suddenly, bam! We're stabilized, hanging tautly under the parachute.

We jettison the thermal shield that ensured we didn't burn up when we re-entered the atmosphere; our windows were blacked over from the heat, but now an extra layer of covering peels off and we can see the blue morning sky. All remaining fuel has already been vented to ensure we don't burst into flames when we hit the ground.

We try to catch our breath, weak after the multi-axis disorienting tumble, the wildest of amusement park rides. To complete the effect, our seats suddenly slam upward, rising automatically to the top level of their shock absorbers to cushion us from the brunt of what's about to happen. The crush of acceleration helps us tighten our straps. We know the moment of impact will be bad; the seats' liners were custom-built to mold to our bodies so that our backs don't break. Just before impact no one says anything, not even Roman, who's been narrating our descent as he is supposed to, talking a mile a minute the whole way down, telling the ground what's going on. We're all clenching our teeth, lightly, so we don't bite through our tongues.

Our little gamma-ray altimeter waits for an echo from the ground, and then, two seconds before impact, sends a command to fire our optimistically named Soft Landing Rockets—gunpowder charges that cut our descent rate to 5 feet per second. They turn a horrific car crash into a survivable one: we hit the hard ground of Kazakhstan, a ton of steel, titanium and human flesh. It's windy on the steppe, so our chute drags us over onto our side like a chopped tree, and we roll end over end a few times until Roman flicks a switch to cut the parachute lines, and we . . . stop. The Soyuz rests on its side. I'm upside down, hanging heavily in my straps from the ceiling, stunned, shaken, stirred.

A normal landing, right on target: we hear the drone of the

search and rescue helicopters. We inhale the burnt, acrid smell of our spaceship. Tom points to the window: where moments before there had been space, now there is pale brown, powdery dirt. We hear a jabber of voices—the Russian ground crew.

We're back on Earth, at last.

* * *

Next thing you know the hatch is being pried open and there's blue sky, bright sunshine, the smell of fresh air and living things, a commotion of voices. Arms reach in to lift Roman out of the capsule. Someone else digs out the samples and science, the things that need to be put in a freezer or on a plane right away. Tom is carried out next, then it's my turn. I was NASA's rep at several landings, so the ground crew knows me, and the guy who lifts me out says, in Russian, "Chris, the clip is magnificent, it made us proud." He's talking about "Space Oddity," I realize, and he means he's proud of this business we're both in. It's a nice way to be welcomed back when you've fallen from the sky.

I'm pale and blinking after months without sunlight, and so weak and rubber-limbed that I need to be carried over and propped up in a canvas chair beside Tom and Roman, who is already joking with the medical staff and looking great, like he's ready to play a round of golf. I am not. Doctors and nurses are wiping the dirt off my forehead; I accidentally touched the charred edge of the Soyuz while getting out and then touched my face, so I look as though I've been smeared with charcoal. They're asking if I'm all right, tenderly, and covering me with a blanket. NASA and CSA officials, local dignitaries and Russian soldiers are buzzing around. It's overwhelming, after being with no more than five other human beings for the past five months, to be surrounded by a crowd of well-wishers, especially after the physical excesses of crashing down to Earth.

My helmet comes off and someone hands me a satellite phone. Helene. A few reporters press forward for the photo op: E.T. calls home. I hear my wife's voice, sure and clear, relieved and happy. I tell her I love her, then ask the question: Did the Leafs win the game? No, she tells me, they're out of the play-offs. They'd gone down in flames, just like me.

I'm smiling, doing my best to impersonate a person who doesn't feel disoriented and sick. But my arms feel so heavy I can barely lift them, and I stay motionless, to reduce exertion. Every part of my body feels sore or shocked, or both. It's like being a newborn, this sudden sensory overload of noise, color, smells and gravity after months of quietly floating, encased in relative calm and isolation. No wonder babies cry in protest when they're born.

After sitting still for 15 minutes, and handing over my personal belongings to a support person who will make sure they don't mysteriously disappear (anything that's flown in space is a collector's item), I'm carried, chair and all, into a hastily erected medical tent to be transferred to a cot. By this point I'm retching, feeling just terrible. Medical staff clean me up, help me out of my Sokhol and my Leafs shirt, now soaked with sweat, and into my regular blue flight suit, then put in an IV to give me more fluids so I don't faint.

Next, along with Roman and Tom, I'm loaded into an armored vehicle, a long, low-ceilinged thing that reeks of diesel fumes, to be carted a few hundred yards to a helicopter. Not a peak experience when you're nauseated. We each get our own MI8, a Russian military transport helicopter with a bed, nurse, support person and doctor. I'm most interested in the bed. I'm dazed, and every time I move my head I feel like I'm spinning through space and time. I fall asleep almost immediately.

Landing at the airport in Karaganda about an hour later, I'm at least refreshed and strong enough to sign the vehicle's door (one astronaut or cosmonaut did it once, a spur-of-the-moment impulse that was instantly institutionalized as a must-do—and it is kind of cool to add your own signature to those of colleagues you know personally or by reputation). Tom, Roman and I are helped into a car and whisked off to a ceremony where a local VIP presents each of us with a purple robe and black hat that look a bit like something Merlin might wear, and a two-stringed gourd-shaped guitar. Young Kazakh women in formal dress provide standard offerings for travelers: salt, bread and water.

Then there's a press conference and the first question is, "Did you know that 'Space Oddity' has had seven million hits?" I didn't, actually. The number sounds unbelievable and I'm really feeling sick now, but need to explain that Expedition 34/35 was not about a music video. Rather, the purpose of the music video was to make the rare and beautiful experience of space flight more accessible. I babble something in Russian about the importance of having human beings in space, not robots, then some merciful person trundles me off to the bathroom, where I can be sick without worrying about bad press.

Later, we're driven back to the airport's taxiway, where Roman gets on a plane to Russia, and Tom and I board a NASA G3, a small jet with two beds in the back and room for 10 passengers. Farewells are bleary and to the point, not sentimental. We don't have it in us. We're all ready to sink into the oblivion of sleep. It takes about 20 hours to return to Houston, and between naps, medical staff monitor our vital signs and clamor for more blood and urine samples; NASA is trying to get as much data as possible on the physiological impact of long duration space flight. While the jet refuels in Prestwick, I have a shower, sitting on a chair. It feels amazing to wash my hair, to be clean all over for the first time in nearly half a year.

When I get off the plane in Houston, bone-tired and not yet steady on my feet, a small group is there to greet me. I kiss Helene, hold her for a moment. Being able to talk to her without a two-second delay, as we had on the ISS phone, feels like both a decadent luxury and a familiar comfort. Family and friends have come, people I know and like and have thought of over the past five months, and I take a bit of time with each of them. It's both pleasant and slightly stiff, like a receiving line at a wedding—a necessary ceremony marking a transition. Helene is watching, knowing I want to leave, so we go, straight to crew quarters.

It is 11:30 at night, which means, time to give 14 vials of blood then do a few sims and tests to assess our balance and ability to concentrate! Tom and I had always known we'd have to do this and also knew it was important, but of course, given the hour and how we were feeling, we felt a little grumpy about it, especially when we realized we were bombing the tests.

There was a hand-eye coordination test, similar to one I'd done 21 years earlier in Ottawa during astronaut selection: alternately using your left hand, then your right, then both, you stick pegs into a row of holes on a peg board, being evaluated for speed and accuracy. It's like a cribbage drag race. I was clumsy after zero gravity, and had trouble grabbing just a single peg from the shallow bin without sending the rest of them flying to the ground. Then there was a computer test, where you had to try to keep a cursor inside a circle that was moving all over the screen, while simultaneously typing in numbers that showed up on another screen. The worst, though, was the motion simulator. You sit in a small round cockpit mounted on a tilting platform, responding to computer images that simulate flying a NASA T-38, driving a race car on a winding mountain track and maneuvering a bulbous rover on Mars. Even pre-flight, the visuals were provocative, but now the experience was truly sickening.

I don't think I've ever been as happy to go to bed as I was that night. After months of being able to somersault effortlessly through the air, I could barely hold my head up. Bed was about all I could handle.

But I was happy that evening for another reason: I felt we'd succeeded at something difficult. Expedition 34/35 had been a success scientifically, and social media had made it an educational success, too. I knew I would never return to space; I'd finally achieved a goal I'd devoted most of my life to achieving. I didn't feel sad about that. I felt elated: I'd done it! And I knew there was more to do, even if, at that moment, I wasn't quite sure what, exactly. But if seeing 16 sunrises a day and all of Earth's variety steadily on display for five months had taught me anything, it was that there are always more challenges and opportunities out there than time to experience them.

Yes, we bashed into the ground pretty hard in Kazakhstan. But I didn't view it as the end of something. Rather, I saw it as a new beginning. And in that sense, at least, it was a soft landing.

13

CLIMBING DOWN THE LADDER

WHEN THE SHUTTLE WAS IN SERVICE, I used to fly a small plane between Houston and Cape Canaveral pretty regularly. It wasn't a scenic route: civilian aircraft are supposed to avoid military airfields, and there are many in that part of the world, so I had to fly directly above the interstate most of the way. I followed along I-10 like any commuter, only 10,000 feet up, so I could see more of the gray ribbon of road that stretches across the flat, sandy Gulf Coast states. Nothing too exciting.

But one time, flying in a twin-engine Beechcraft Baron with Russ Wilson, a friend of mine who's a firefighter, I was coming over the Panhandle when something brushed lightly against my leg—my bare leg: it was a blistering summer day, so we were wearing shorts. Figuring it was probably an electrical cord dangling underneath the pilot's seat, I shifted around in my seat to get away from it. A moment later, though, there it was again, touching my leg. Weird. I looked down and what I saw, rising up from the floor, was a black snake. Not a garter snake and not a python, either, but certainly the biggest reptile I've ever seen in a cockpit. I instinctively jerked my feet up onto my seat, which caused Russ to look down and spot the snake.

For a few long seconds we stayed like that, frozen in disbelief.

If a flight has been particularly challenging, fighter pilots like to say they've been busy "killing snakes and putting out fires." But this really was a snake, and trying to kill it at 10,000 feet seemed ill-advised. A failed attempt on its life was not going to make it any more kindly disposed toward us. Russ didn't wait: he grabbed the clipboard that held our checklists and used it to pin the snake down on the floor. Then he grasped the thing in the approved manner, just behind its head, and yanked it out from under my seat.

Which was the snake's cue to start whipping the rest of its body around, frantically trying to escape, while I tried to keep flying the plane as though nothing at all unusual was happening. What next?

Without a whole lot of discussion, we decided to open the window on my side. It was small, just big enough to let smoke escape from the cockpit if there was a fire, but we were going 200 miles per hour, so suddenly it was like we were in the middle of a hurricane. The noise was wicked, our ears were popping from the drop in cabin pressure, there was an ornery serpent lashing all over the place. But firefighters are good in a crisis. Russ calmly leaned over me, stuck his hand out the window and somehow forced most of the snake out there too, then let go. Poof. It was gone. We quickly closed the window, and then we thought to look around: Any more snakes in here? How did that one get in, anyway? Did that really just happen?

A blast of nervous, post-adrenaline laughter. Already the episode felt unbelievable and had acquired the sheen of an anecdote. Then I wondered, "Where is that snake now?" When I pictured the scene below—a black snake writhing in free fall, confused and disoriented, smashing onto the windshield of a car—I stopped laughing, because I had a pretty good idea how that would feel.

Coming back to Earth from space, I felt as though I was being rudely flung down from the heavens, and then, splat! An hour before, I'd had the powers of a superhero—I could *fly*. Now I was so weak I could barely hobble around unassisted. My body, spoiled by the luxury of weightlessness, aggressively protested the return to gravity. I was nauseated and exhausted; my limbs felt leaden, my coordination was shot.

And I was a bit irritable when, at post-flight press conferences, a reporter invariably asked how I felt "now that it's all over." In fact, it wasn't over: every flight is followed by months of rehabilitation, medical testing and exhaustive debriefing with everyone from the top administrators at NASA to the people who resupply the ISS. But the reason the question bothered me was the implication that space flight was the last worthwhile experience I'd ever have, and sadly, from here on in, it was all downhill. I don't look at myself or the world that way. I view each mission as just one thread in the overall fabric of my life—which is, I hope, nowhere near over.

* * *

If you start thinking that only your biggest and shiniest moments count, you're setting yourself up to feel like a failure most of the time. Personally, I'd rather feel good most of the time, so to me everything counts: the small moments, the medium ones, the successes that make the papers and also the ones that no one knows about but me. The challenge is avoiding being derailed by the big, shiny moments that turn other people's heads. You have to figure out for yourself how to enjoy and celebrate them, and then move on.

Astronauts who've just returned from space get a lot of help from NASA with the "moving on" part. When you report back to the Astronaut Office at JSC, there's no hero's welcome. Rather, you get a brisk acknowledgment—"Good job"—before being unceremoniously booted off the top rung of the organizational ladder, at least in terms of visibility and prestige. Astronauts fresh off the Soyuz are reabsorbed back into the support team as middle-of-the-pack players, essential but not glorified.

In most lines of work there's a steady, linear ascent up a well-defined career ladder, but astronauts continuously move up and down, rotating through different roles and ranks. From an organizational standpoint, this makes sense: it keeps the space program strong at all levels and also reinforces everyone's commitment to teamwork in pursuit of a common goal—pushing the envelope of human knowledge and capability—that's much bigger than we are as individuals. For astronauts, too, it makes sense, because it helps us come right back down to Earth and focus on our job, which is to support and promote human space exploration. Any inclination we might have to preen is nipped in the bud, because our status has changed overnight and we are expected to deliver in a new, less visible role, not sit around reminiscing about the good old days when we were in space.

At NASA it's just a given that today's star will be tomorrow's stagehand, toiling behind the scenes in relative obscurity. For instance, Peggy Whitson, who was Chief Astronaut and ran the office in Houston for three years, is now back in the regular pool of astronauts, supporting other astronauts in orbit and hoping for an assignment with no better chances of being selected than anyone else has. One thing that makes this kind of transition easier is that the line between being a member of a crew and a member of the office is already more blurry than might be readily evident to outsiders. A CAPCOM, for instance, does some training and goes to sims with a crew, then supports them or is on call every day of their flight, and afterward, also attends debriefs. In a very real way, then, the CAPCOM is integral to that crew—as is the entire cast of people who directly support any mission.

If you're part of that support team, you know full well that the meaning and significance of your work isn't determined by how visible it is to outsiders. And once you've stood on the top rung of the ladder, where you are fully aware of how critically important the people on the ground are to the success of your mission, it's actually easier and more meaningful, in some respects, to support other astronauts on their missions.

But I'm not going to pretend that a flat organizational structure has no drawbacks or that giving up a dream assignment is a thoroughly joyous experience. Even Pollyanna would have some mixed feelings about it. However, astronauts get so much practice swapping between lead and supporting roles that it does get easier over time.

And sooner or later you realize that it's better for everyone, including you, if you climb down the ladder graciously. After being the Director of Operations for NASA in Russia for a few years, when I went back to Star City to train I sometimes found myself wondering, "Why is the new DOR doing X that way?" I quickly learned that as the ex-whatever, you only get so many golden opportunities to keep your mouth shut, and you should take advantage of every single one. I wasn't in charge anymore. My role was limited to observing and—only if it seemed absolutely necessary—trying to mold the process through subtle means. Usually it wasn't necessary. Frequently, the "issue" was simply that the other person's managerial style was not the same as mine.

Even if you've been a plus one in a certain role—maybe especially if you've been a plus one—once your stint is over, it's time to aim to be a zero again. This turns out to be easier than you might think right after you get back from space. At least at first, you feel so crummy, physically, that zero looks like a big step up.

* * *

The rule of thumb is that you need a day on Earth to recover from each day in space, and happily, that proved true after my first two missions. They were relatively short—8 days in 1995, 11 in 2001—so I had a few rough days right after we got back, but a week or so later, I was back to normal.

Returning from Expedition 34/35 was different. After five months in space my body hadn't just adapted to zero gravity, it had developed a whole new set of habits. After a few steps my feet, no longer accustomed to bearing weight, felt as though I'd been walking across hot coals. Sitting down didn't bring much relief: now my feet felt exactly the way I imagine they'd feel if someone had pounded them repeatedly with a mallet. Plus, seated, I was uncomfortably aware of my tailbone; when you're used to resting on air, weightlessly, sitting on a chair, weightily, really does not feel good. But neither does standing. After elongating in space, my spine was now compressing again, so my lower back was constantly sore. I was surprised how long it took for these side effects to go away. Months later, my feet and back were still complaining—frequently and loudly—about what a drag gravity is.

My heart also developed new habits in space. By the time I returned to Earth, it had forgotten how to pump blood all the way up to my head, so simply standing up required it to work strenuously. After a few minutes on my feet, my heart rate edged up to

130. Meanwhile, my blood pressure was dropping, so I felt faint. To help my circulation, I wore a g-suit for a few days to keep steady pressure on my calves, thighs and gut. It's a lot like squeezing the bottom of a balloon to force air upward; the g-suit doesn't hurt, it just feels like something heavy is pressing on your lower body. But even so, I felt incredibly dizzy if I stood up quickly, which made me wary of bathrooms; in the first few days post-flight, there's a real danger of keeling over and cracking your head open on a tiled floor (one astronaut I know did pass out when he got up to pee). That's why, in post-flight quarantine, there's a chair in the shower in crew quarters. Although the vertigo became less acute, I continued to experience it for a long time and learned to stand still after I got up and let the dizziness pass before attempting something rash, like walking across the living room.

Part of the problem was that my vestibular system—the mechanism in the inner ear that controls balance—was totally bewildered post-flight. On the ISS, it got used to responding only to my body's own rotations and accelerations, because up was down and down was up. Back on Earth, though, gravity was suddenly pulling me down and the floor was now holding me up, trapping my inner ear in what felt like a constant acceleration that, inexplicably, my eyes couldn't perceive. It's extremely nauseating, worse than the most sickening ride at the fair. My body reacted as though the symptoms were being caused by a neural poison, and urged me both to purge it and to lie down, so that I'd metabolize the poison more slowly. I took anti-nausea meds on and off for about 10 days after landing; sometimes I felt just fine, but other times, I looked and felt green.

My stomach recovered faster than my sense of balance. At first, walking was difficult, a drunk's stagger, but as I re-adapted I got better at it (so long as I kept my eyes wide open). Still, for at least the first week, I over-corrected, swinging wide on turns, bumping into things and tilting forward as though I was walking into gale-force winds. All of this meant it wasn't safe to drive for a couple of weeks, which was just fine with me, because I was profoundly, almost unbelievably, tired, like an invalid recovering from a debilitating illness.

I slept heavily and peacefully, which was an unexpected plus; for the first few days after my Shuttle flights, I'd had the weird sensation that I was floating above my bed (I'd been away such a short time, my body was probably thoroughly confused). This time, I had no such trouble. My bed was where I felt most comfortable, physically, and I craved sleep so much that I was sneaking several catnaps a day.

Fortunately, NASA has top-notch personal trainers who work with us and our doctors from initial assignment through recovery: Astronaut Strength, Conditioning and Rehabilitation specialists (ASCRs). My first day back in Houston, they asked me to lift my arms over my head and then to lie down on the ground and try to lift up my legs. I could do both things but just barely. Lying on the mat, I felt as though two people were sitting on top of me, pinning me to the floor. After the empowering environment of space, where I could move a refrigerator with one fingertip, it seemed . . . well, unfair. Despite exercising two hours a day on the ISS, I was, back on Earth, a weakling.

A lot of what happens to the human body in space is really similar to what happens during the aging process. In post-flight quarantine, in fact, Tom and I tottered around like two old duffers, getting a preview of what life might be like if we made it to 90. Our blood vessels had hardened; our cardiovascular systems had changed. We had shed calcium and minerals in space, so our bones were weaker; so were our muscles, because for 22 hours a day, they'd encountered no resistance whatsoever.

On the plus side, with the help of the rehab specialists, we could reverse most of the damage, and in the meantime, doctors could poke and prod us to gain insight into physical changes related to aging. For the first few months back, astronauts are essentially outsized lab rats.

We even run mazes, of a sort. Scientists want to know more about the aftereffects of long-duration space flight, so they repeatedly administer the types of tests we were given that first night in quarantine, as well as a few new ones. There was, for instance, a form of hopscotch: a long rope ladder was laid out on the ground and I had to jump, hop and skip the length of it, in patterns ranging from ones you'd see on the playground to *Saturday Night Fever*–type moves. There were timed sprints too, where I had to weave around cones while running forward, backward and sideways. I'd done all of this pre-flight too, so my baseline results could be compared to my post-flight scores. Not surprisingly, my agility and reaction times were, in the first few weeks after landing, quite a bit less impressive.

Other tests were more involved. For one that measured how our competence was affected by our circadian rhythms, two plastic things that looked like bolts were taped onto my forehead and chest while an arm cuff monitored my vitals. I had to go out for a burger one evening looking like Frankenstein. For a balance test, I was first wired with sensors and trussed up in a harness, then asked to stand on a small platform looking at a picture of the horizon. The scientists had me tip my head backward and forward while they moved the horizon scene or the small platform to see whether I'd lose my balance. Or my lunch. (I came close.)

Between all the tests, debriefs and media interviews, there was very little downtime. I felt slightly detached and observational; consciously engaging seemed like work. I felt oddly withdrawn.

I was back on Earth, but not back to my life on Earth. For the first month or so, I was at JSC most of the day, even on weekends. Helene packed healthy lunches and drove me back and forth until, after three weeks, my doctor agreed that it was safe for me to get behind the wheel.

Working back up to a normal workout took a lot longer. I spent two hours a day with the rehab specialists, who eased me back into exercise using equipment such as a floating treadmill: I wore a pair of rubber shorts that zipped into a big rubber balloon—by inflating it, they could control how much of my weight my legs needed to support while running. I started with about 60 percent, which matched the pull-down force that the shoulder and waist bungees had provided on orbit.

After two months, when I finally got the okay to go for a run outside, my legs felt heavy and slow, and I could feel my insides sloshing around as I awkwardly, clumsily pounded along. My cardiovascular response, too, was disappointing: my feet were still apparently top priority for blood flow, and my forgetful veins and arteries were still in no big rush to pump anything up to my lungs and head. I realized that for at least six months I simply wouldn't be able to do activities that might involve sudden cardiovascular demands, like water-skiing or team sports. Aside from anything else, my bones couldn't handle any shocks or stresses. After returning from the ISS, one astronaut had an innocuous fall that nevertheless resulted in a broken hip—I didn't want to add to that particular database.

* * *

About three weeks after landing, Tom and I headed back to Star City together for the traditional official Russian debrief and ceremony. For him, it was the last leg of the multi-year journey that was Expedition 34/35. For me, it was the last leg of a 21-year career as an astronaut. Months earlier I'd told the CSA that I was retiring, and shortly there would be a public announcement.

The journey to Star City, then, felt both comfortingly familiar and a little strange. Packing, flying, being picked up at Domodedovo airport in Moscow by Ephim, a smiling, slightly devilish long-time friend and driver for NASA—I'd done it all before, many times, yet the knowledge that I might not do it again changed the experience. After Ephim dropped us at the NASA townhouses, my comfortable, no-frills home away from home for so many years, I actually felt . . . free. I hadn't been completely on my own, and in control of my own schedule, for a very long time. Years, maybe. There were no doctors, no family, no trainers, just the pleasantly selfish simplicity of being responsible only for myself. Tom and I both commented on the decadence of it, then happily left one another alone. I walked around the pond, read quietly, caught up on email unhurriedly. It felt . . . enjoyable.

Ephim had told me that Roman had bought a new car, so I was prepared when he turned up in a gold BMW convertible the next morning. It was clearly a reward he'd promised himself in return for all the time away, and an earthly pleasure I understand well: I have two convertibles, an old Thunderbird and a newer Mustang. Roman and I grinned at one another, two middle-aged men unashamed of our predictability.

Our technical debriefs later that day were straightforward and perfunctory, since Roman had already gone over all the details with Roscosmos. The trip was really more an opportunity for the three of us to thank and toast old friends—the instructors and trainers who'd worked with us for years, helping us get ready—as well as a photo op. We were presented to the media for a Q & A in Russian, followed by a lot of smiling and joining of hands to oblige the photographers. There were more journalists there than usual because it was a big day in the history of space flight; we were also celebrating the 50th anniversary of Valentina Tereshkova's flight: she was the first woman in space. Along with my friend Alexei Leonov, the world's first spacewalker, she joined our crew to pose for pictures in front of Yuri Gagarin's statue. I couldn't quite fathom this unlikely juxtaposition of my heroes and my own recent history, and filed it away as one more amazing event to revel in now and try to figure out later.

Our visit culminated in an awards ceremony in a long, too-warm hall filled with instructors, DOR staff, Roman's family, NASA management, Roscosmos and Energia bosses, local politicians and many youth groups. One by one, people came forward to honor our crew with short speeches, handshakes and gifts: plaques, watches, books and endless, enormous bunches of flowers. Tom, Roman and I each had our own table just to hold all this stuff, and by the end the tables were overflowing. Tom and I gave our flowers to the ladies in the DOR office and turned the pricey gifts over to NASA management, who put it all in storage, just as the president does with his costly gifts (government employees can't accept expensive presents, but we were allowed to keep a few small items).

After I'd said my goodbyes and we were sitting in the NASA van, I had a very strong feeling that I would never be back in Star City. Suddenly, I didn't want to leave. I wanted to seek out my friends and colleagues all over again, to hug them once more and relive all the experiences we'd shared. I wanted to do something to elevate my departure from mere leave-taking to big event, which is what it was for me. A chapter in my life was over. But instead, I sat quietly in the van, melancholy as the so-familiar faces and places eased out of view, but mostly grateful. Russia had been good to me.

Back on the other side of the ocean, the mood was jolly, celebratory, hectic. The prime minister of Canada invited me over for a visit. I was the Calgary Stampede's parade grand marshal, an honor especially given the city's Herculean efforts to clean up after a devastating flood in time for the annual celebration. There were parties in Houston and Montreal, where the CSA is headquartered. I shook hands, gave endless interviews, felt stronger every day. I packed up my desk at JSC, and we boxed up everything in our Houston home; we moved back to Canada amidst a swirl of flattering articles and tributes. The "Space Oddity" video had opened doors, musically, and I performed for large crowds at big events. There were so many requests, I had to come up with a form letter politely declining speaking engagements and endorsement offers. It was exciting. It was exhausting.

And it was, I knew, ephemeral.

* * *

Who was vice president three administrations ago? Which movie won Best Picture at the Oscars five years ago? Who won gold in speed skating at the last Olympics? I used to know. These were big events at the time, but soon afterward, they were largely remembered only by the participants themselves.

A space mission is the same. The blast of glory that attends launch and landing doesn't last long. The spotlight moves on, and astronauts need to, too. If you can't, you'll wind up hobbled by self-importance or by the fear that nothing else you do will ever measure up.

Some astronauts do end up mired in the quicksand of bygone celebrity, but they are the exceptions. More than 500 people have had the opportunity to see our planet from afar, and for most of them, the experience seems to have either reinforced or induced humility. The shimmering, dancing show of the northern and southern lights; the gorgeous blues of the shallow reefs fanning out around the Bahamas; the huge, angry froth stirred up around the focused eye of a hurricane—seeing the whole world shifts your perspective radically. It's not only awe-inspiring but profoundly humbling. Certainly it drove home to me how near-sighted it would be to place too much importance on my own 53-odd years on the planet. I take great pride in what our crew accomplished while we were on the ISS, especially the record amount of science we completed and the fact that Tom and Chris Cassidy pulled off an emergency spacewalk. But in the annals of space exploration, we'll be lucky to merit a footnote.

This is not to say that space travel has made me feel irrelevant. In fact, it's made me feel I have a personal obligation to be a good steward of our planet and to educate others about what's happening to it. From space, you can see the deforestation in Madagascar, how all that red soil that was once held in place by natural vegetation is now just pouring into the ocean; you can see how the shoreline of the Aral Sea has moved dozens of miles as water has been diverted for agriculture, so that what used to be lake bottom is now bleak desert. You can also see that Earth is a durable, absorbent, self-correcting, life-supporting place that has its own problems—natural ones, like ash-spewing volcanoes. But we make matters infinitely worse through poor stewardship. We need to take a longer-term view of the environment and try to make things better wherever we can.

I feel a sense of mission about this that I didn't have before I went to space, and people who know me sometimes find it exasperating. Recently a friend got frustrated with me because while we were out for a walk, I kept stopping to pick up trash, which slowed our progress considerably. This turns out to be one of the little-known aftereffects of space flight: I now pick gum wrappers up off the street.

Understanding my place in the grand scheme of the universe has helped me keep my own successes in perspective, but it hasn't made me so modest that I can no longer bear applause. I bear it just fine, and actually get a kick out of the hoopla around launch and landing. Still, I also know that most people, including me, tend to applaud the wrong things: the showy, dramatic record-setting sprint rather than the years of dogged preparation or the unwavering grace displayed during a string of losses. Applause, then, never bore much relation to the reality of my life as an astronaut, which was not all about, or even mostly about, flying around in space.

It was really about making the most of my time here on Earth.

Some people assume that after going to space, everyday life on Earth must seem mundane, lackluster even. But for me, the opposite has been true. Post-flight, I feel the way you might feel after a really interesting trip you'd been planning and anticipating for years: fulfilled and energized, as well as inspired to see the world a little differently.

A high-octane experience only enriches the rest of your life—unless, of course, you are only able to experience joy and feel a sense of purpose at the very top of the ladder, in which case, climbing down would be a big comedown. Suddenly, there's no more applause, and you're facing the stark reality of having to take out the trash and deal with the imperfections of daily life.

The whole process of becoming an astronaut helped me understand that what really matters is not the value someone else assigns to a task but how I personally feel while performing it. That's why, during the 11 years I was grounded, I loved my life. Of course I wanted to go back to space—who wouldn't?—but I got real fulfillment and pleasure from small victories, like doing something well in the Neutral Buoyancy Lab or figuring out how to fix a problem with my car. If I'd defined success very narrowly, limiting it to peak, high-visibility experiences, I would have felt very unsuccessful and unhappy during those years. Life is just a lot better if you feel you're having 10 wins a day rather than a win every 10 years or so.

One of the accomplishments I'm proudest of has nothing to do with flying in space or even being an astronaut: in 2007, my neighbor Bob and I built a dock at the cottage. A decade-old disagreement with the previous owners of our cottage had led to two exactly parallel, increasingly dilapidated docks, bizarrely separated by a 1-inch Do Not Cross zone that exerted a strange, magnetic pull on my aging dog's foot. Bob and I set out with the manly goal of making some minor repairs ourselves, to save a bit of money. Our loving wives made frowny faces, raised their eyebrows and asked, "Are you kidding?"

Thus inspired, we decided to raze both docks and start over, welding together a single, mighty superstructure you could land a small plane on. All we had to do was buy out a lumberyard, rent a barge, hire a pile driver and labor from dawn to dusk throughout our summer holiday. As with building the docking module for Mir, we were solving a long-term problem and uniting two previously warring parties, and the experience was just as rewarding and satisfying—maybe even more so, because the task was self-appointed and completion depended solely on our own skills and ingenuity. Building that dock felt like the best job in the world, and I still view it as the crowning achievement of that year, when, by the way, I was also NASA's Chief of Space Station Operations in the Astronaut Office.

The truth is that I find every day fulfilling, whether I'm on the planet or off it. I work hard at whatever I'm doing, whether it's fixing a bilge pump in my boat or learning to play a new song on the guitar. And I find satisfaction in small things, like playing Scrabble online with my daughter, Kristin—we always have a game going—or reading a letter from a first grader who wants to be an astronaut, or picking gum wrappers up off the street. Because of all of this, plus the fact that at NASA I got so much experience climbing down the ladder, I wasn't afraid to retire.

Endings don't have to be emotionally wrenching if you believe you did a good job and you're prepared to let go. When the Shuttle program was winding down, reporters repeatedly urged me to go public with my private pain: "We know you're sad about the end of the program, but just *how* sad are you?" I wasn't sad at all. I was extremely proud. I was part of a team that flew the Shuttle 135 times and used it to put the Hubble telescope into orbit, to build part of Mir and to help build the ISS. Along the way, we recovered from two devastating accidents, the *Challenger* and *Columbia* disasters. After *Columbia*, so many people said it was time to mothball the Shuttle—what was the purpose of going to space again, why risk lives? But somehow, despite the media's simplistic focus and all the naysayers who had no knowledge of the issues but plenty of opinions, we prevailed and the Shuttle flew again, safely. The complexity of the project we needed the Shuttle for was astonishing—the Station's design wasn't even complete when the first pieces of the ISS launched—yet we did it. So there's no reason at all to be sad that the Shuttle era is over and the spaceships are in museums. They were great workhorses of space exploration, and they served their purpose.

I view my own retirement the same way. I did the best I could and I served my purpose, but the time has come to move on. Unlike the Shuttle, however, I am not destined for a museum, and as it turns out, that's my own fault. Several years ago, a museum in British Columbia wanted a plaster cast of my face to place on a dummy (insert witty comment here). Along with the instructions in the package, they sent a helpful note that said, "It's not rocket science." So Helene and I cracked open the kit. It had green goop for my hair, eyebrows and moustache, pink goop to spread everywhere else on my face and plaster strips to hold it all together. But despite a thorough team briefing, it was pretty much a disaster. Helene plastered over my nostrils, so we nearly had a fatality. The goop set too quickly and the plaster didn't stick to the goop. The mask crumbled. And after lying on the floor in a pool of chalky mud, I got an ear infection.

I decided not to attempt a redo, recognizing that perhaps this was not meant to be. Anyway, a faceless dummy is actually the perfect symbolic representation of one of the most important lessons I've learned as an astronaut: to value the wisdom of humility, as well as the sense of perspective it gives you.

That's what will help me climb down the ladder. And it won't hurt if I decide to climb up a new one, either.

ACKNOWLEDGMENTS

✳

THIS BOOK HAS BEEN REALLY IMPORTANT TO ME. The process of writing it helped cohere many disparate memories, thoughts and events not only on paper but within myself. To be able to hold it in my hand, a tangible product of my choices in life, is akin to a birth, and feels near miraculous.

More importantly, however, I am grateful for the way it has aligned so many people, friends and family, in working toward a common goal. Many of them don't even know how much they contributed, as they were consulted only in my memories and in self-reflection on how they helped shape my beliefs.

There is no way to thank everyone, as the caboose would outweigh the train, so I cherry-pick those most precious and accept the inevitability of forgetting someone key and dear.

My family gave me unending support, stories, guidance and permission. I have dedicated the book to Helene, as there is no one more deserving and beloved. To Kyle, Evan and Kristin, who grew up with a passionate, focused, regimented and largely absent dad, the Colonel says thank you. I am hugely proud of each of you, and brag on you to everyone. My parents, Roger and Eleanor, passed their values on to me and trusted me with them, especially when I insisted on pursuing the vesper that became the reality of space flight. The soaring heights of this life found their foundation in you. Brother Dave—who traveled all the way to Baikonur just so we could play together one last time before launch—your music is with me everywhere and always.

Good people often select themselves. Rick Broadhead, your joy, understanding and tenacity make you a formidable friend and agent. Elinor Fillion, I have appreciated your practical assistance and moral support every step of the way. Kate Fillion, you have been through my words so many times, you truly, scarily know me. It's an intimacy that conductors must have with new scores, seeing and hearing the sound of the music before the first note is ever played. You are legend.

Anne Collins and John Parsley, your bravery and trust are only matched by your patient relentlessness to get it right.

Finally to each of you, unmentioned, who have shared a piece of this life, those who are long gone, the exquisitely precious friends who have shaped my days and continue to do so, I salute and thank you. It's been a self-pinching ride to this point. Hugs for everybody.

INDEX

CHRIS HADFIELD ANSWERS YOUR FAQS

What part of astronaut training was the most fun?

No contest: spacewalk training. Astronaut preparation can often seem endless—classroom theory, one-on-one spacecraft systems lectures where your attention can never lapse, Russian language lessons, physical fitness, homework night after night, for years. The rare day we get to spend underwater, practicing our space-walking skills, is different. We train in a huge, deep swimming pool—so big that almost the entire International Space Station can be submerged in it—using the buoyancy of water to simulate weightlessness. To be wearing the spacewalking suit (or EMU—Extravehicular Mobility Unit), surrounded by the very realistic mock-up of the Space Station, able to float and move and work as if the whole thing was in orbit, makes a day of training the closest we can get to the reality of spacewalking. It truly makes you feel like you're an astronaut.

Does food taste different in space than it does on Earth?

Being in zero gravity is confusing for the body. It not only affects your balance system (the inner ear takes a holiday as your eyes take over telling you which way is up), but it also removes the perpetual weight that pulls the fluids inside your body down toward your feet. When that lets go, you feel like you're suddenly being sucked up toward the ceiling. After a few days you adapt; your legs get visibly skinnier and your face fatter, gorged with blood, like you're endlessly standing on your head. Your sinuses don't drain nearly as well, your head pounds and your tongue feels swollen. It's as if you have a perpetual head cold: your nose and taste buds are not at their best.

As a result food tastes blander. Something that may have been deliciously nuanced on Earth now tastes boring and unremarkable. To spice it up we squirt pepper onto it (suspended in olive oil in a little plastic squeeze bottle), as well as liberally add hot sauce and wasabi. One of my favorite foods on orbit is dehydrated shrimp cocktail—not so much for the shrimp as the horseradish in the cocktail sauce. It startles the sinuses and brings welcome tears to your eyes (though without the effect of gravity, tears don't fall; they just stay in your eyes and slowly evaporate).

What do the ISS and space smell like?

When I first boarded a space station I expected it to smell bad, like a locker room or a cave or a basement. But it had almost no smell at all. The air purification system was so strong, and the environment so clean and humidity-controlled, it smelled more like an office building. I could detect a slight trace of oil and electricity, my nose reminding me that I was inside a machine, but just a trace. And I never picked up even a whiff of body odor from the other astronauts, though we lived in fairly close quarters with limited bathing capability (we only take sponge baths).

Space itself would be hard to smell. We only go outside completely cocooned in our pressurized spacesuits because going naked into the vacuum of outer space would kill us. We may get a hint of what space might smell like just after a spacewalk, when we pop off our helmets and sniff. The air in the airlock carries a distinctive, lingering odor, a bit like the smell just after someone fires a gun, or maybe when fat drips onto the coals in a barbecue. It's likely not the smell of space at all, but space's vacuum pulling the last traces of volatile particles out of the metal and plastic walls of the airlock.

Regardless, I find it delightful that the after-odor of space smells like cordite or brimstone, as if an unseen witch was just there.

Did you see and feel the impact of meteorites on the ISS?

Yes, because we're living inside a thin-skinned metal can. You often hear pops and twangs. Sometimes it's just metal expanding and contracting in the direct sunshine, but occasionally it's more like a ricochet—a stray bullet from the universe that (luckily) didn't pierce the skin. The first time you hear that particular sound, or notice a small scar on a window or solar array from a previous impact, it's a bit unsettling—a vivid reminder that you're living in a dangerous place. But like experiencing tremblers in an earthquake zone, after a while you accept that the small ones are okay.

What was your scariest moment in space?

No one wants a frightened astronaut. If things are going poorly on board a spaceship, it's not the astronaut's role to hide under a table, but to take action and make things better. No one else can. That's why our training takes so long. Not only do we have to learn what to do in all circumstances, but we need to practice for every eventuality so many times that they are no longer frightening to us.

I did feel raw fear once, though. We were soaring high over Australia in the dark, and I was looking for the lights of the cities along the coast, when I saw a shooting star. It was a big one, streaking across the Outback for a few seconds, and I remember thinking just how beautiful it was. But a moment later I realized that a big, dumb rock had just raced past us and hit the Earth's atmosphere like a bomb. If something that size had hit our spaceship, the damage would have been instant and extensive and fatal. The station would have depressurized like a pop can exploding and then each of us would have violently suffocated, tumbling in the vacuum of space. I felt a shiver of fear run all the way up my back, with a surge of adrenaline following it. It gave a new meaning to the idea of wishing upon a falling star.

Is it comfortable to sleep on the ISS?

It is a whole new type of comfortable to sleep in weightlessness. Even on the most expensive Earth mattress you occasionally have to roll over or adjust your pillow. In orbit, you can relax every muscle in your body. At bedtime you float into your sleeping bag (which is loosely tethered to the wall with a couple of shoelaces), do up the long zipper and shut off the light. Because there's no effect of gravity pushing you into your mattress, you are perfectly relaxed and your whole body can go deliciously limp. Your arms and leg joints bend a bit and float up, your neck slumps forward like a napping passenger on a plane; every muscle takes a rest. You can feel the slow pulse of your heartbeat, moving you slightly against the nothingness. When space travel eventually becomes inexpensive enough, it may well be the 'space sleep spa' that attracts the largest crowd.

What time zone do you live by in space? Do you switch off the lights at 'night'?

The International Space Station circles Earth at tremendous speed—about 17,500 mph, or 8 km/sec—which is necessary to stay in orbit. We make a complete turn around the Earth every 90 minutes, or 16 times per day. That equates to 16 sunrises and sunsets per day, a bizarre and beautiful new reality. It also changes the concept of local time, making wristwatch-setting completely arbitrary.

Early in the space program, for everyone's convenience, astronauts would just stay on the same time zone as Mission Control: Moscow time for cosmonauts and Houston time for NASA. But for the International Space Station, with 15 nations working together worldwide, a compromise was necessary. We settled on Greenwich Mean Time, same as the clocks in London, England, more accurately known as Coordinated Universal Time. Thus the crew members of ISS get up and go to sleep in the same time zone as the Queen in Buckingham Palace.

As commander, I shut off most of the ship's internal lights at bedtime to help make it feel more natural. I was concerned about a sudden emergency in the middle of the night, so left enough lights on that we could see what we were doing, but no more than that. It became a pleasant ritual for me to put the ship to bed at night and brighten it up in the morning.

What was the favorite thing you witnessed in space (captured in a photo or not)?

The Space Station was soaring over the Indian Ocean and across Indonesia on a very stormy night. I was floating in the Cupola window; far below me I could see a vast expanse of thunderstorms flashing on the coming horizon, like we were approaching a horde of paparazzi. The closer we got, the brighter the lightning bolts became, until the white explosions were directly beneath us. Just then my crewmate Tom Marshburn floated into the Cupola too, a rare chance for two astronauts to steal time together and look at the world.

I hadn't realized what lightning truly does until that day. Each explosion of electricity not only sends a bolt to the ground, but lights up the entire thundercloud like a giant flashbulb. And lightning is contagious. A flash would start at one end of the clouds and travel the length of them, as if some giant hand with a huge white highlighter was emphasizing the entire storm, end-to-end. Tom and I gasped and laughed and pointed in awe and delight.

The storm was gigantic, and it took us minutes to fly its length. The combination of the visual spectacle and sharing it with Tom made it the favorite thing I witnessed in my 2,597 orbits of our planet.

What was your favorite city to look at from space?

Pyongyang, North Korea, because I know so little about it. I have traveled extensively, visiting over 60 countries, but certain parts of the world are still a mystery to me and pique my curiosity. The effects of the political regime in North Korea are visible from space—a country cloaked in darkness at night, showing perceptibly the lack of human development. The brilliance of nearby Seoul and South Korea are a stark contrast. Even with the longest camera lens, Pyongyang looked unimpressive, a small, dark grey blob with a sports stadium barely visible. I looked every time we passed over, and still I learned very little about it. It increased my empathy for the people living in its darkness.

How did your time as commander of the ISS change your perspective on our planet?
Each of us is raised with a sense of 'us and them.' Initially the 'us' is just family, and everyone else is 'them.' As we get older and more experienced, more and more people join the 'us' but there is usually still a 'them.' Even when I launched on my third space flight, at the senior astronaut age of 53 and having lived in a lot of places, I still carried a distinct, though murkily defined, personal idea of 'us and them.'

Once in orbit, though, with time to not only work but to gaze at the world over a period of months, I noticed my perception shifting. As I sent pictures to the ground and commented on them, I found myself unthinkingly referring to everyone as 'us.' It came inexorably from witnessing how we live: the repeated pattern of human existence right across the planet. I would see a city that I knew well and just 30 minutes later see that exact same pattern of settlement in a city I had never heard of. It forced me to face the commonality of the human experience, and our shared hopes and desires.

We are all 'us': crewmates on the same big ship, working and hoping for a little joy, some grace and better opportunities for our children. All of us.

What do you see as the next frontier in space exploration?
The impulse to explore is as old as humanity itself, predating written history. The evidence is in everywhere we live; by our best reckoning, it took about 70,000 years for us to explore and settle all corners of the Earth, on foot and by boat. A hundred years ago, it was just barely possible to explore Antarctica; now technology enables thousands to visit it each year, and nearly a hundred to live year-round at the South Pole itself. Fifty years ago our first

explorers ventured into space; since November 2000, astronauts from 15 of the world's leading countries have been living permanently off the planet on the ISS.

We are all explorers; it's our nature. Because of this enduring human urge, I think that space exploration will naturally continue. It's not a race, and it's not for entertainment purposes. It's just the inevitable human desire to better understand the universe around us, and to seek increased opportunities for a better life.

The obvious next frontier is the Moon. It's only a three-day voyage with our current engine technology, and it presents many of the challenges that we have to solve in order to be able to explore farther: power generation, resupply, in-situ resource use, radiation protection, communication, navigation, closed-cycle environments, psychological support, surface transportation, propulsion and so on.

A Moon colony would provide a large, stable platform for observing Earth and the rest of the universe with no atmosphere in the way, and if we made a serious error in judgment, we'd have the chance to evacuate back to Earth.

One small step at a time—it's our heritage.

Do you think humans will ever travel to Mars and eventually colonize it?

Of course I do. The same question has been asked by different people about different places throughout history. In 1491, for example: 'Do you think humans will ever cross the ocean and eventually colonize what they find?' Or in 1900: 'Do you think humans will ever travel to Antarctica and even live there?' We have always explored to the limits of our ability and technology, and then moved to the places we found interesting and/or beneficial. Often our first attempts have failed, both during the probing phase (Shackleton, Earhart, Apollo 1) and the colonization phase (L'Anse aux Meadows, the Jamestown Settlement, Skylab). We have sent probes throughout the solar system, including six manned ventures to the surface of the Moon, and have begun settlement in Earth orbit on space stations. Eventually we will travel as far as Mars, and beyond. We are only limited by our ability to invent and persevere.

Would you go to Mars if you had the chance?

Yes, I would love the chance to go to Mars, but there are so many technologies that we have to invent and then test before we are able to go. No sane person is going to fund or climb aboard a doomed mission to Mars. What rocket lifts us away from Earth, for one? What is the spacecraft that can sustain interplanetary life for six months with no turning back, featuring completely reliable propulsion, food supply, water recycling and navigation, while its occupants withstand radiation, even madness? How do we slow down at Mars and enter its atmosphere for a survivable landing? What habitat would solve all the challenges of sustaining life indefinitely on the surface? And how would we reverse the entire process to return to Earth, or would the trip be one-way?

Most critically, why go? What are we doing there that makes it worthwhile? We'll only launch such a mission when our technology decreases the risk (and the cost) to an acceptable level to match the benefits to humanity.

Saying yes, I'd go to Mars, is the easy part.

If you hadn't become an astronaut, what career path would you have chosen?

I decided to turn myself into an astronaut when I was only nine, but even then I recognized that it was near impossible and I needed a Plan B. People get sick, things go wrong, and then where are you?

All along the path, I chose to pursue activities that led generally in the direction of space flight, but that were interesting and suited me on their own. I was a downhill ski racer and instructor, and learned to love controlling speed and hazard through skill. I studied engineering, and enjoyed the complexity and challenge of problem-solving and invention. I was an Air Force pilot and have a fundamental love of flying, and could have become an airline pilot like my father. But it was when I trained as a test pilot that I really found my niche: speed, complexity and hazardous situations overcome through skill and methodical problem-solving, giving a result that makes flying safer and better understood for all. If the Canadian Space Agency had phoned me that fateful Saturday and said, 'Sorry, thanks for applying, we've chosen someone else,' my plan was to go back to school, get a Ph.D. and then teach at one of the universities or other organizations that have flying and flight test programs. And maybe do a little skiing.

Which aspects of Earth life did you miss the most when you were on the ISS for 146 days?

Most people think that space flight must be lonely, being so far away from home and locked in a spaceship with a small crew for a long time. Space flight has even become a metaphor for loneliness in popular songs like 'Rocket Man' and 'Space Oddity.' In truth, though, I was never lonely while in space. I was actively working with all my crewmates daily, had near-constant radio contact with mission control centers in many countries, and the whole world was in my window, all 7 billion of us there to see and wave at. And I could phone my family pretty much whenever I wanted and had a private videoconference with my wife weekly.

Thinking about it in orbit, though, it occurred to me that I had never met a lonely farmer or mountain climber. The loneliest people I know live in cities, surrounded by noise and people and hubbub. I don't think it's location that makes people feel lonely—it's individual psychology. The same is true for what we miss, and how homesick we get. In general I try to spend as little time missing things as possible. It seems a shame to me, when surrounded by life and potential adventure, to be wishing I were somewhere else. What I did miss during the five months I spent off the planet was simple physical contact: the random ability to give someone a hug or kiss a loved one. Even with the nausea and weakness I felt after landing, it was great once more to be back-slapping, hugging and kissing happily in the tumult of Earth-bound life.

Which aspects of space life do you miss the most? Do you dream about space?

I don't miss space life, but not because it wasn't a tremendous experience. It's because I lived it fully at the time it was happening, and feel immensely satisfied with our crew's level of result. Dwelling on the past is backwards to how life proceeds. I view my space flights as amazing things that have led me to where I am now, giving me skills and perspectives that allow me to appreciate the present and be ready for the future.

I do dream of space flight, sometimes, but I don't believe those dreams carry any great significance. I view them as the idle meanderings of the reorganizing mind, and seldom remember them with any clarity. I'm much more a planner than a dreamer.

Do you believe intelligent life exists elsewhere in the universe? Five hundred years ago, the commonly held theory was that the Earth was the center of the universe, and everything revolved around us. To the uneducated eye, it sure looked that way—the sun and Moon rose on one side of the horizon, traveled across the sky, and set on the other side. It also fit nicely with the human tendency to self-importance.

As new technology was invented, though—namely Galileo's telescope—we could see that we were wrong, and it was, in fact, the Earth turning that made the sky go around. As technology improved, we could see evermore clearly that not only are we not the center of the universe, but just one tiny spinning planet orbiting a pipsqueak star on the outer edges of an unremarkable galaxy, among hundreds of billions of galaxies.

Even better technology has recently started to show us planets around other stars. Now we know how many planets orbit the average star, and can estimate the total number of planets that exist. More interestingly, we can predict how many of them resemble Earth in size and temperature, and as a result might have developed life as we know it. The number is staggering: in our galaxy alone, we think there are about 10 billion Earth-like planets orbiting stars like our Sun. Given that there are hundreds of billions of galaxies, the odds are strong that life has evolved somewhere else. To think that we are the only life in the universe is just an extension of the same arrogance that made us think we were the center of it all.

Why spend money in space when people are hungry on Earth? In all societies, we need to balance how we spend our money. The vast majority needs to be on human health and services. A portion also needs to go to education. In addition, some needs to be for research and exploration. It is vital that we take care of our people, educate our young and develop opportunities for the future. But if each country doesn't challenge its citizens with demanding ideas and possibilities, they will either go elsewhere, which is a loss, or not realize their potential, which is a tragedy. The key is to decide on the right balance of needs and budget. When you look at the actual figures, I think the space agencies of the world get it about right. I know the Canadian Space Agency works very hard with the money they are given to do as well for our country as they can—developing useful products, better understanding the world and human health, and inspiring our next generation.

What is your advice to a young person who wants to become an astronaut?

Let's turn this question around: if you were hiring astronauts, what capabilities would you look for? I've been involved in several astronaut recruitments, and the key characteristics we need are:

- the proven ability to learn,
- the proven capability to make good decisions, and
- physical fitness.

For a young person it's possible to turn these desired traits into plans and actions, to improve the odds of someday flying in space:

- the simplest proof of learning capability is an advanced degree in a complex field—a master's or Ph.D. in physics or math or engineering or medicine, etc.
- there are careers that demonstrate good decision-making, especially where the consequences matter. That's why the space agencies hire test pilots and doctors, and people who

have undertaken intense technical programs. Start making decisions today, and stick with them. Decision-making is a skill, and the more you do it, the better you get at it.

- most astronaut applicants are eliminated due to medical reasons. To be healthy enough to risk six months on a space station not only requires normal, exercise-based fitness, but also fundamental congenital health. It's important to take care of your body, and also to get a thorough physical to understand your personal medical status.

The other key piece to consider is what else will make you an interesting and useful crewmate? Learn to fly, to scuba dive, to speak other languages; travel the world and pursue opportunities to serve others. All these things count.

Finally, accept that the odds of becoming an astronaut are terrible. Never make it your personal measure of success or failure. That's setting yourself up for disaster. Rather, make it a long-term goal, unlikely but still agenda-setting, which will help guide you in your daily decisions and small victories. That's where your life truly will be lived.

What's next for you in your life on Earth as a retired astronaut?
I see retirement from the astronaut corps as a natural step along my particular winding road of life. Just as I used to be a downhill ski racer and instructor, farmer, fighter pilot and test pilot, I used to be an astronaut. To have succeeded as an astronaut required intense focus, but now I welcome the chance for different pursuits. I'd like to study archeology, one of my first loves. I want to teach, and have taken a part-time professorship at the University of Waterloo in Ontario, as well as working with elementary and high school students. I like mental challenge and problem-

solving, and as a result consult to the aerospace industry. I enjoy music, and am writing songs and working with other artists to perform and record, including a full orchestral version of the suite of tunes I wrote in orbit with my brother Dave and son Evan.

Mostly, I want to be useful, to feel that not only have I contributed to what is important to me, but that I will continue to be able to do so.